AF869570

DNA Transposable Elements Research

DNA Transposable Elements Research

Kaito Yoshida and Miyu Aoki
Editors

Nova Biomedical Books
New York

Copyright © 2008 by Nova Science Publishers, Inc.

All rights reserved. No part of this book may be reproduced, stored in a retrieval system or transmitted in any form or by any means: electronic, electrostatic, magnetic, tape, mechanical photocopying, recording or otherwise without the written permission of the Publisher.

For permission to use material from this book please contact us:
Telephone 631-231-7269; Fax 631-231-8175
Web Site: http://www.novapublishers.com

NOTICE TO THE READER

The Publisher has taken reasonable care in the preparation of this book, but makes no expressed or implied warranty of any kind and assumes no responsibility for any errors or omissions. No liability is assumed for incidental or consequential damages in connection with or arising out of information contained in this book. The Publisher shall not be liable for any special, consequential, or exemplary damages resulting, in whole or in part, from the readers' use of, or reliance upon, this material.

Independent verification should be sought for any data, advice or recommendations contained in this book. In addition, no responsibility is assumed by the publisher for any injury and/or damage to persons or property arising from any methods, products, instructions, ideas or otherwise contained in this publication.

This publication is designed to provide accurate and authoritative information with regard to the subject matter covered herein. It is sold with the clear understanding that the Publisher is not engaged in rendering legal or any other professional services. If legal or any other expert assistance is required, the services of a competent person should be sought. FROM A DECLARATION OF PARTICIPANTS JOINTLY ADOPTED BY A COMMITTEE OF THE AMERICAN BAR ASSOCIATION AND A COMMITTEE OF PUBLISHERS.

Library of Congress Cataloging-in-Publication Data

DNA transposable elements research / Kaito Yoshida and Miyu Aoki (editors).
Includes bibliographical references and index.
ISBN 978-1-60456-531-7(hardcover)
1. Transposons. I. Yoshida,Kaito.II. Aoki,Miyu.
[DNLM: 1. DNA Transposable Elements. QU 470 D629 2008]
QU452.3.D63 2008
572.8'69—dc22

2008011189

Published by Nova Science Publishers, Inc. New York

Contents

Preface

Transposons are segments of DNA that can relocate (transpose) to different positions within the genome of a single cell. In the years since their intial discovery in 1948 by Barbara McClintock, these mobile genetic elements have come to be widely recognized as ubiquitous components of genomes representing all major branches of life; furthermore, transposons have been developed into powerful tools for molecular biology and, in particular, functional genomes, in wide range of organisms. More recently, transposons have been developed into a technology platform for vertebrate genetics with application areas including gene therapy, transgenesis, somatic mutagenesis (cancer research), and germ line mutagenesis for gene discovery. This book presents new and important research from around the world in this field.

Chapter I - Transposons, or more colloquially "jumping genes", are segments of DNA that can relocate (transpose) to different positions within the genome of a single cell. In the years since their initial discovery in 1948 by Barbara McClintock, these mobile genetic elements have come to be widely recognized as ubiquitous components of genomes representing all major branches of life; furthermore, transposons have been developed into powerful tools for molecular biology, and, in particular, functional genomics, in a wide range of organisms. In bacteria, for example, the Tn*10*-derived T-POP transposon has served as an effective system for target gene overexpression, enabling genome-wide screens for regulatory genes of interest. Conversely, transposon mutagenesis also serves as a classic method for gene disruption; for example, transposon-based signature-tagged mutagenesis (STM) has been used to generate loss-of-function alleles on a large scale, identifying essential and infection-related genes in microbial pathogens. In the budding yeast, multipurpose transposons have been applied on a genome-wide scale to generate insertion libraries facilitating the analysis of gene disruption phenotypes, hypomorphic phenotypes, and protein localization. In *A. thaliana* and *D. melanogaster*, *Ac*/*Ds* and P-elements, respectively, have been utilized as tools for insertional mutagenesis. More recently, transposons have been developed into a technology platform for vertebrate genetics with application areas including gene therapy, transgenesis, somatic mutagenesis (cancer research), and germ line mutagenesis for gene discovery. While far from comprehensive, the transposon-based approaches summarized in this chapter exemplify the utility of transposable elements as powerful tools for functional genomics, while hinting at even broader applications in the near future.

Chapter II - Bacterial transposons encoding a variety of phenotypic traits are widely disseminated in diverse species, thus allowing the host strains to expand their niches. Catabolic transposons responsible for the degradation of man-made and/or natural organic compounds provide the host cells to use these compounds as the sources of carbon, energy, and/or nitrogen, and have been identified worldwide in various bacterial strains from the environments contaminated with herbicides, pesticides, petroleum, etc. Studies on these transposons have revealed that the catabolic genes for the compounds are often located on the class II (Tn*3*-like) transposons, many of which are known to carry resistance genes for antibiotics or heavy metals. In contrast to the insertion sequence (IS)-composite transposons, the class II transposons are able to transpose at high frequencies even when they are large (>50 kb) in sizes, allowing the wide dissemination of large gene clusters for the degradation of aromatic compounds such as toluene/xylenes, naphthalene, and carbazole. In the last two decades, many class II catabolic transposons have been identified and clarified their structures, molecular functions, and in addition the entire sequences of large self-transmissible plasmids carrying the transposons. In this review, the authors summarize the recent information on bacterial class II catabolic transposons, including their transposition mechanisms. Also described are the roles of these elements in establishment and evolution of the large catabolic plasmids.

Chapter III – Transpositions have been identified as built-in responses to genome shock [17] and as agents of genome variation. In addition, they seem to give rise to several universal genome features that may be essential to distinguish functional genomes from meaningless poly-nucleotide and, thus, may facilitate horizontal genome transfer. Examples of such universals appear to be the universal strand symmetry, sometimes called "Chargaff's second parity rule", and the almost universal tri-nucleotide frequency distribution, called "majority profile". Both could be explained as products of the actions of countless inversions and inverted transpositions in the evolutionary past of the genomes [1-3].

As reported here, these particular products of inversions/transpositions give rise to yet another apparent universal in the following way. The genomes whose triplet profiles are almost identical with the majority profile define a large group of organisms called the 'majority group'. In spite of the fact that the triplet profiles of their genomes are almost identical with the majority profile, they may locally deviate from it, nevertheless. In this chapter, the author introduces the graphs of these local deviations as "profile signatures" and shows their high degree of conservation during the evolution of orders. In addition, they offer rapid and comprehensive overviews over chromosome rearrangements and areas of synteny.

The signatures themselves are obviously not genome universals. However, the underlying process which causes their local deviations from the majority profile appears to be yet another universal, which can be linked directly to the action of inversions /transpositions.

Chapter IV - Apart from well-known bacterial methods of mutation, the horizontal transfer of genetic material within and between microbial genera plays crucial role in the emergence of novel antibiotic resistance genes and their dissemination observed worldwide. Rapid and widespread appearance of multidrug resistance and similar resistance patterns observed in closely - as well as distantly - related bacterial species are due to presence and activity of mobile genetic elements, especially integrons. Integrons are able to capture and exchange genes in a small mobile elements called cassettes in a process of site-specific recombination. They have been assigned to at least four distinct classes based on *int* integrase gene homology. Class 1 integrons are most prevalent in clinical isolates and usually associated with multidrug resistance phenotypes. This chapter describes PCR-based detection method of class 1 integrons and characterization of captured gene cassettes in clinical isolates of different serovars of *Salmonella enterica* in Slovakia during 2002-2006. In order to gain complex picture, the authors have developed a new microarray-based method for molecular characterization of *Salmonella* isolates in terms of integrase and antimicrobial resistance genes detection. They developed the procedure using DNA microarray for localization of integrons as well as embedded gene cassettes carried either on chromosomal DNA or plasmids thus providing valuable insight to dissemination process and epidemiology of emerging resistance genes.

Chapter V - Insertion sequences are regarded as the simplest mobile genetic elements, encoding only the information necessary to enable and regulate their transposition. Bacterial insertion sequences display high diversity and may be categorised into multiple families based on related transposase domains, open reading frame structures, terminal inverted repeats (IRs), and similarities in the outcome of the target site DNA upon integration. Although originally considered to be passive genetic elements or 'junk' DNA, many years of accumulated evidence has now conclusively demonstrated that they can act as important regulators of gene expression and mediators of genome plasticity. Thus, despite their structural simplicity, the genotypic and subsequent phenotypic effects of bacterial insertion sequences may be complex. This complexity is enhanced by the fact that many bacterial insertion sequences are located on extrachromosomal DNA elements that may be transmitted between species and have the capacity to integrate into host chromosomes, carrying accessory genes with them. Typically, the transposition of an insertion sequence results in a negative effect on gene expression in the target DNA due to the disruption of coding or promoter sequences. However, many insertion sequences have been shown to possess the ability, through several related but discrete mechanisms, to upregulate genes downstream from the insertion sequence integration site. Here the authors review some of the phenotypic consequences of insertion sequence-mediated gene activation, focussing on its effects in human bacterial pathogens. In particular, they discuss how insertion sequence-mediated gene activation can alter bacterial virulence and lead to antibiotic resistance.

Chapter VI - Ascidians, which are commonly known as "sea squirts", belong to phylum Chordata with cephalochordates and vertebrates. As this phylogenetic position suggests, ascidians share characteristics with vertebrates, such as dorsally located neural tubes and notochords. A cosmopolitan species of ascidian, *Ciona intestinalis*, is commonly used as a material for developmental and genome biology. The draft genome sequences of *Ciona* have revealed that this ascidian possesses a compact genome with little redundancy of genes. The *Ciona* life cycle is approximately two to three months, and inland culture systems have been developed. Thus, *Ciona* provides an excellent material on which to perform genetic analyses. Recent studies have shown that a Tc1/*mariner* superfamily transposon *Minos* causes both excision and transposition in *Ciona*. Activity of *Minos* is sufficient to cause germline transformation. So far, techniques such as creation of marker transgenic lines, enhancer detection and insertional mutagenesis have been introduced in *Ciona* with the *Minos* system. *Minos* inserted in the *Ciona* genome can be remobilized by providing transposase in the transgenic lines and, utilizing the remobilization technique, efficient enhancer detection screening has been established. In this article, these recent technological achievements of *Ciona* genetics based on the *Minos* transposon are reviewed.

Chapter VII - CRISPRs (clustered, regularly interspaced short palindromic repeats) are loci present in many bacterial genomes and characterized by alternating minisatellite direct repeats (DR) and non-repeats (spacers). In the complete genome sequence of *Corynebacterium diphtheriae* strain NCTC13129, the authors previously identified two CRISPR loci 39 kb downstream and 180 kb upstream of the *oriC* region, that they named DRA and DRB, respectively. Here, they analysed these loci in silico in more detail. A search against complete GenBank revealed a homology of seven spacers in both loci (one DRA spacer and six DRB spacers). Noteworthy, four of them showed such homology with either phage hypothetical proteins, or plasmid transposon encoded genes related to integration or recombination. This result corroborates with recent observations made on other bacterial genomes, suggesting that CRISPRs are structures able to take up pieces of foreign DNA as part of a defence mechanism against its invasion. In *C. diphtheriae*, five of the seven found homologies are on minus-strand sequences in GenBank hinting at the RNA interference as a mode of action in CRISPR loci. CRISPR analysis also revealed striking points of similarity between *C. diphtheriae* and other *Corynebacterium* (*C. jeikeium*) partly owing to the horizontal gene transfer. In particular, the last spacer in the *C. diphtheriae* DRA locus showed homology to the direct repeat motif in the large 60-unit CRISPR locus in *C. jeikeium*. *C. jeikeium* is a skin pathogen and thus likely a more ancient species than human respiratory pathogen *C. diphtheriae*. Accordingly, it seems more plausible that this spacer was acquired from *C. jeikeium* CRISPR locus and not *vice versa*. Analysis of the regions flanking both CRISPR loci in *C. diphtheriae* suggests that at least one them (DRA), functionally or structurally, may be prone to acquire DNA fragments, acting as “insertion preferential locus” (*ipl*).

In: DNA Transposable Elements Research
Editors: K. Yoshida, M. Aoki, pp. 1-22
ISBN: 978-1-60456-531-7
© 2008 Nova Science Publishers, Inc.

Chapter I

Might as Well Jump: Transposons as Tools for Functional Genomics

Rui Jin and Anuj Kumar*
Department of Molecular, Cellular and Developmental Biology
and Life Sciences Institute, University of Michigan, Ann Arbor, MI 48109-2216, USA

Abstract

Transposons, or more colloquially "jumping genes", are segments of DNA that can relocate (transpose) to different positions within the genome of a single cell. In the years since their initial discovery in 1948 by Barbara McClintock, these mobile genetic elements have come to be widely recognized as ubiquitous components of genomes representing all major branches of life; furthermore, transposons have been developed into powerful tools for molecular biology, and, in particular, functional genomics, in a wide range of organisms. In bacteria, for example, the Tn*10*-derived T-POP transposon has served as an effective system for target gene overexpression, enabling genome-wide screens for regulatory genes of interest. Conversely, transposon mutagenesis also serves as a classic method for gene disruption; for example, transposon-based signature-tagged mutagenesis (STM) has been used to generate loss-of-function alleles on a large scale, identifying essential and infection-related genes in microbial pathogens. In the budding yeast, multipurpose transposons have been applied on a genome-wide scale to generate insertion libraries facilitating the analysis of gene disruption phenotypes, hypomorphic phenotypes, and protein localization. In *A. thaliana* and *D. melanogaster*, *Ac/Ds* and P-elements, respectively, have been utilized as tools for insertional mutagenesis. More recently, transposons have been developed into a technology platform for vertebrate genetics with application areas including gene therapy, transgenesis, somatic mutagenesis (cancer research), and germ line mutagenesis for gene discovery. While far from comprehensive, the transposon-based approaches summarized in this chapter exemplify the utility of transposable elements as powerful tools for functional genomics, while hinting at even broader applications in the near future.

Introduction

First discovered in the mid-1940s by Barbara McClintock, transposons are ubiquitous DNA elements capable of physically relocating within a host genome [1]. Typically, this process of relocation, or transposition, is accomplished through the action of several proteins encoded within the transposable element itself. These proteins enable transposition by one of two paths: the transposable element may be duplicated and this new copy of DNA sequence inserted elsewhere in the genome, or the existing element may be excised and reinserted at a new site. The mechanism of this transposition is linked to the structure of the transposable element, and these structures and mechanisms have been investigated extensively [2]. This sizable interest in transposon structure and function stems from the evolutionary significance of transposons in shaping genomes and from the phenotypic consequences of transposition. Dating back to early studies in bacteria, transposons have been recognized as media of transfer for genetic material within and between species [2]. Furthermore, transposons are, of course, mutagens; transposition can result in serious inherited damage and disease in host cell genomes. In fact, numerous diseases have been linked with transposon insertions, including breast cancer, Ewing's sarcoma, type II diabetes, hemophilia A/B, severe combined immunodeficiency, and predisposition to cancer [3].

Beginning in the 1970's, however, transposons began to assume a different and equally important role in molecular biology as a powerful source of laboratory tools and reagents. We present here an overview of these transposon-based tools, summarizing many applications of transposon technology towards biological studies in both prokaryotes and eukaryotes. In general, these applications highlight a straightforward theme: for use as a biological tool, the transposon itself can be modified to permit user-directed control of transposition and convenient identification of insertion products. Transposons are very amenable to genetic manipulation; the terminal ends of transposons are typically required for transposition, but the central components of transposable elements can be modified extensively by the addition of organism-specific markers and reporter genes. The specific nature of these modifications is dependent on the application, with insertional mutagenesis and gene/promoter trapping as the most common laboratory uses for transposon technology. In these applications, modified transposons are introduced into a genome of interest, either *in vitro* or *in vivo*, and subsequent mutants assayed for a given readout by some convenient system, often utilizing a transposon-encoded reporter. Specific examples of this generalized approach are presented in this chapter, with relevance to organisms ranging from bacteria to humans.

Relative to other mutagenic approaches, such as X-ray irradiation, chemical-mediated mutagenesis, and PCR-directed gene replacement, transposon-based systems provide several advantages. First and foremost, transposon mutagenesis is cost-effective. Although initial investments in labor and time are necessary to devise an appropriate transposon system for a given study, the transposition reaction itself is typically straightforward and inexpensive. Thus, a large collection of insertion mutants can be generated easily by this approach, with fewer concerns regarding off-site mutations, such as chromosomal rearrangements from X-ray irradiation. In addition, transposon mutagenesis is convenient, allowing for easy identification of insertion sites following transposition. Classically, time-consuming and labor-intensive processes, such as positional cloning, are necessary in order to identify affected genes and mutation sites generated by x-ray irradiation or chemical treatment. In contrast, transposon insertion sites can be readily identified in mutants of interest by virtue of the transposon sequence itself; this sequence serves as a tag for amplification of flanking genomic DNA by inverse PCR / vectorette PCR [4]. DNA microarrays may provide an even more attractive approach for the identification of transposon insertion sites. Recently, several groups have developed strategies by which regions of genomic DNA adjacent to insertions may be mapped onto whole-genome microarrays. For example, Winterberg and Reznikoff [5] used T7 promoters located at the ends of a modified bacterial transposon to generate RNA transcripts representing chromosomal DNA flanking the transposon insertions; these RNA transcripts were then labeled and hybridized against a whole-genome oligonucleotide microarray, thereby identifying insertion sites present within a selected population of mutants. Similarly, Wheelan *et al.* [6] devised a transposon insertion site-profiling chip (TIP-chip) upon which selectively amplified transposon flanking regions can be mapped.

Transposon-based approaches, however, do possess distinct limitations. In particular, transposition is not strictly random; transposons exhibit biases in insertion site selection, making it difficult to obtain insertions affecting all genes in a genome. As a result, transposon insertion screens typically fall short of saturation. Bacterial transposons are generally less biased in insertion site preference than eukaryotic transposons; even so, the prokaryotic transposon Tn*3* displays detectable insertion specificity [7], and it is currently unclear if any transposon-based system can yield a saturating phenotypic screen. In addition, the phenotypic analysis of insertion mutants can sometimes be complicated, since it is not always clear whether a given insertion allele represents a null, hypomorphic, or misexpression mutant. Obviously, this point is not a concern when mutations are generated by approaches such as directed start codon-stop codon gene replacement, for example.

On balance, however, the advantages of transposon-mediated mutagenesis outweigh its drawbacks, and transposon systems have become favorable genetic tools for functional genomics in a wide host of organisms.

Transposon-Based Applications in Microbial Genomics

Signature-Tagged Mutagenesis

Transposon-based approaches have been applied with significant success to the study of pathogenic bacteria; signature-tagged mutagenesis (STM) is one such important application, enabling the identification of hundreds of new virulence determinant genes in a broad range of bacterial pathogens. STM represents an evolution of traditional transposon mutagenesis, wherein transposons are marked with unique "signature tags", enabling individual insertion mutants to be resolved within a mixed mutant pool (Figure 1).

Hensel *et al.* [8] presented a classic application of STM, performing a high-throughput parallel genetic screen for virulence determinant genes in *Salmonella typhimurium*. In this work, the authors designed a pool of signature tags, with each tag containing a unique 40bp DNA sequence flanked by two 20bp invariable arms.

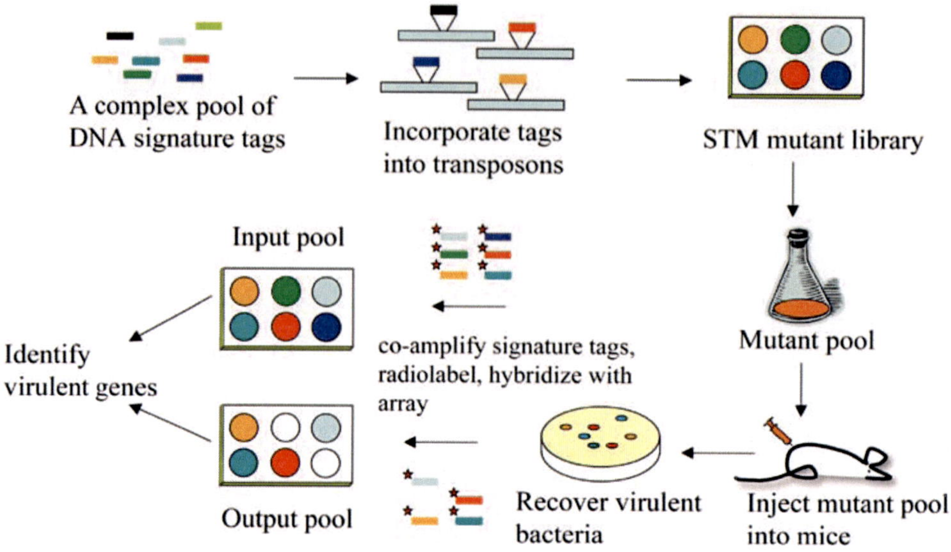

Figure 1. Signature-tagged Mutagenesis. Overview of signature-tagged mutagenesis and its application towards the identification of virulence determinant genes in bacterial pathogens. Signature tags are not drawn to scale in the illustration above.

By this design, the tags could be co-amplified by virtue of the common arms, while still being differentiated because of the central unique sequence. These tags were incorporated into transposons for mutagenesis of bacteria, and the transposon-mutagenized bacterial strains were subsequently assembled into a signature-tagged mutant library. This mutant pool (the input pool) was used to infect mice, and virulent bacteria were recovered from the infected tissues. Tags from virulent bacterial DNA (the output pool) were subsequently amplified, radiolabelled and hybridized with probe DNA from the input pool. DNA present in the input pool but absent or less abundant in the output pool represents mutants with attenuated virulence. Using STM, Hensel and co-workers discovered a SPI-2-encoded type 3 secretion (T3S) system, which transfers over 15 virulence proteins from bacteria across their phagosomal membranes into host cells [8]. This drives formation of a specialized compartment named the *Salmonell*a-containing vacuole, required for intracellular replication of *Salmonella* during infection. Recently, the SPI2 T3S mutant strain has become a promising vaccine for typhoid fever, the disease resulting from *Salmonella typhi* infection.

Compared to classical methods of site-directed gene inactivation, STM offers two major advantages. First, application of STM by the approach described above represents a form of negative selection, allowing for the discovery of virulence genes without preconceptions derived from circumstantial evidence or those inferred from the apparent pathophysiology of the disease. Second, STM bypasses a major limitation of classic genetic screens in animal models by enabling the parallel analysis of many mutants in a single test. Although STM is a powerful tool for the study of pathogenic bacteria, it is not without its limitations. Most notably, in pooled samples, interference between mutants is often observed, yielding complimentary effects that can complicate analysis of results. In addition, STM suffers from technical problems, such as a loss of reproducibility and the identification of false-positive results stemming from poor signal quality [8]. This limitation, however, can be overcome by employing multiple rounds of selection. Finally, as with other transpson-based approaches, STM results can be biased by non-random transposon insertion site selection and by the chosen infection model itself.

Despite some technical constraints, STM has been successfully applied towards many functional genomic studies in pathogenic bacteria. *Mycobacterium tuberculosis*, for example, has been a popular target for analysis by STM, as two million deaths annually are reportedly caused by *M. tuberculosis*-related lung infections [9-10]. In the late-1990s, two separate STM studies were undertaken to screen for insertion mutants affected in their multiplication within the lungs of mice. These studies identified two mutated loci related to the biosynthesis (*fadD28*) and export (*mmpL7*) of a complex cell wall associated lipid, phthiocerol dimycocerosate (DIM). DIM has been found to be essential for survival of the pathogen in the lung [8], and DIM has now become an attractive drug target for *M. Tuberculosis* infection therapy.

The T-Pop Transposon System

Conditional expression alleles can be very informative as a means of defining functions associated with essential genes, and transposon mutagenesis provides an effective avenue towards the generation of these mutants on a large scale. Application of the transposon-based T-POP system [11-13] is a case in point. Developed in the late 1990s, the T-POP system has been used successfully to isolate transposon-mutagenized microbial mutants exhibiting desired tetracycline-dependent conditional phenotypes [12]. T-POP is a derivative of the transposon Tn*10*dTet. Its central region includes two tetracycline-resistance genes, tetA and tetR, transcribed by two *tet* promoters, respectively. In T-POP, unlike in the original Tn*10*d transposon, mRNA transcripts initiated at the *tet* promoters do not terminate within the transposon. Instead, they continue into the adjacent chromosomal DNA, leading to tetracycline-regulated induction of adjacent gene expression. The insertion of a typical transposon at a site between an essential gene and its upstream promoter will separate the two, silencing expression of the essential gene and causing cell death; however, transcriptional activity from the *tet* promoters in T-POP serve to bypass this lethality by artificially inducing transcription of the downstream essential gene, generating a tetracycline-dependent conditional mutant. Studies of essential genes using this T-POP system have been implemented in *Salmonella typhi.* In one such study, Hidalgo *et al.* screened 1700 backcrossed T-POP insertion mutants for tetracycline-dependent growth phenotypes and found four mutants capable of growing only in the presence of tetracycline [12]. Among these mutants, one contained an insertion upstream of a tRNA synthetase gene, two were found to be upstream of homologs of other known essential genes, and one was found upstream of the uncharacterized ygqF gene. The T-POP system can also be used to define terminal phenotypes associated with such insertion mutants by simply observing the phenotype of mutant cells transferred to medium lacking tetracycline.

Along similar lines, Judson and Mekalanos [14] described a *mariner*-based mini-transposon, TnAraOut, carrying an arabinose-inducible promoter, oriented such that transposon insertion results in arabionose-dependent expression of adjacent DNA. This transposon is, therefore, suited to identify insertions affecting essential genes by monitoring for growth phenotypes upon removal of arabinose. By this approach, Judson and Mekalanos screened for essential genes in *V. cholerae*, identifying several such genes, at least one of which was specific to the organism.

Transpson Mutagenesis in *Pseudomonas Aeruginosa*

Transposon-based mutagenesis has been used extensively and successfully to screen for desired phenotypes in the opportunistic pathogen *Pseudomonas aeruginosa* [15]. In particular, the Manoil laboratory constructed an extensive insertional mutant library in *P. aeruginosa* by transposon mutagenesis with two different Tn*5* IS50L derivatives, IS*phoA*/hah and IS*lacZ*/hah. The resulting transposon-mutagenized library consists of more than 30,000 unique insertion mutants (15,063 IS*phoA*/hah insertions, 15,037 IS*lacZ*/hah insertions), covering 88% of the total 6.3Mb genome of *P. aeruginosa*. Using this strain collection, Jacobs and colleagues conducted phenotypic screens for the loss of twitching motility and prototrophic growth. In both cases, nearly all previously reported genes contributing to these cell processes were identified, suggesting that the collection offers extensive coverage and a strong platform for additional genomic studies. In total, the utility of this collection is not limited to phenotypic screening. The IS*phoA*/hah and IS*lacZ*/hah transposons possess additional features facilitating downstream studies of gene expression, protein localization, epistasis, and chromosome engineering. For example, *lox*P sites incorporated in the transposons provide a means to remove the selective markers and facilitate construction of double mutants by using another insertion in the collection. Such double mutants may be useful for epistatic analysis or in engineering genomes with deletions between two insertion sites.

Transposon-Based Applications for Functional Genomics in Model Organisms

Genome-Wide Transposon Mutagenesis in Yeast

The budding yeast *Saccharomyces cerevisiae* has classically served as a testing ground for many genomic approaches, and transposon-based technologies present no exception, with large-scale applications of transposon mutagenesis underway in yeast before the sequence of its genome had even been determined. In the mid-1990s, a modified form of the endogenous yeast retrotransposon Ty1 served as the workhorse for a methodology termed "genetic footprinting" [16]. By this approach, a modified Ty1 element is expressed from an inducible promoter, generating a population of yeast strains containing transposon insertions throughout the genome. Insertion mutants are subsequently pooled and grown under desired conditions. PCR is used to amplify a portion of the transposon and flanking DNA from the pooled mutants after growth under selective conditions, and these DNA fragments indicate the relative prevalence of insertion alleles for an individual gene in this population. Genes important for growth under a given condition will be underrepresented in this population, since strains carrying insertions in those genes will suffer from a fitness defect. Thus, genes contributing to yeast cell growth under a given set of conditions can be identified by genetic footprinting.

Biases in Ty1 insertion towards the promoters of genes transcribed by RNA polymerase III [17], however, limit the coverage possible by Ty1 genetic footprinting; instead, bacterial transposons exhibiting less biased transposition have been used more extensively for insertional mutagenesis of the yeast genome (Figure 2A). Bacterial transposons are typically used to mutagenize yeast DNA either *in vivo* in *E. coli* or *in vitro*, and the transposon-mutagenized DNA is then introduced into yeast, an approach termed "shuttle mutagenesis". This form of mutagenesis works particularly well in yeast, since exogenously introduced DNA tends to integrate in the yeast genome almost exclusively by homologous recombination.

Shuttle mutagenesis has been applied extensively in yeast using derivatives of the bacterial transposons Tn*3* and Tn*7* modified for use as multi-purpose transposons [18-21]. These multipurpose transposons have been designed such that a single transposon insertion in yeast genomic DNA can potentially facilitate a variety of functional studies (Figure 2B). This multi-purpose transposon carries a *lacZ* reporter, lacking its initiator methionine codon and upstream promoter sequence. By virtue of this truncated reporter, transposon insertion only results in □-galactosidase activity if the transposon is inserted in a transcriptionally active region of the yeast genome, with the reporter oriented in-frame with yeast protein-coding DNA.

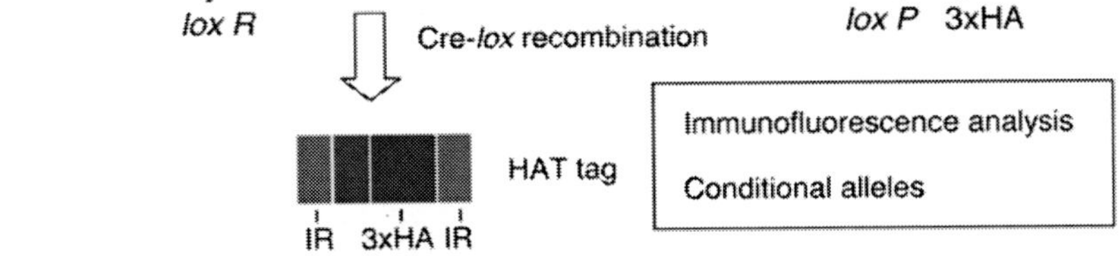

Figure 2. Multi-purpose bacterial transposons for shuttle mutagenesis in yeast. A) Schematic diagram of the bacterial transposon Tn*3*. The left and right terminal inverted repeats are indicated (IR elements) along with the transposase gene *tnpA* and the resolvase *tnpB*. B) Schematic illustration of a Tn*3*-derived multi-purpose transposon. The *tnpA*, *tnpB*, and *bla* genes have been replaced with the 5'-truncated *lacZ* reporter gene, yeast selectable marker *URA3*, and the tetracycline-resistance gene. Two *lox* sites flank this central region, with sequence encoding three copies of the HA tag external to the *loxP* site. This full-length transposon allows for gene disruption upon insertion in the yeast genome. Cre-mediated recombination reduces the transposon to a read-through 93-codon insertion element encoding three

copies of HA; collectively this reduced element is referred to as a HAT tag (transposon-encoded HA tag). HAT-tagged alleles can be used to generate conditional alleles and for protein localization studies by indirect immunofluorescence. Transposon elements are not drawn to scale.

A *lox* site is positioned near each terminus of the transposon, and coding sequence for three copies of the hemagglutinin (HA) epitope lies just external to one of these *lox* sites. After integration of this transposon in the yeast genome, expression of the Cre recombinase can be induced, resulting in excision of the central region of the transposon. Upon Cre-*lox* recombination, the transposon is reduced to a 93-codon sequence encoding three copies of the HA epitope tag. Thus, a single transposon insertion can potentially be used to study gene expression by virtue of the *lacZ* reporter, disruption phenotypes from transposon insertion, and, following Cre-*lox* recombination, protein localization by immunofluorescence using anti-HA antibodies [18].

Michael Snyder's group and others have employed transposon-based shuttle mutagenesis for a variety of functional genomic studies in yeast [19, 22-24]. Ross-Macdonald *et al.* [19] generated expression data for more than 2,000 yeast genes by this approach. In this study, the authors also introduced nearly 8,000 insertion alleles into a haploid strain of yeast and analyzed the resulting insertional mutants for phenotypes under 20 different growth conditions. This analysis identified mutant phenotypes for approximately 400 yeast genes. Kumar and colleagues [21] augmented this insertion collection with mutants generated by *in vitro* shuttle mutagenesis using a Tn7-derived multipurpose transposon; this mutant collection was screened for sensitivity to the anti-cancer drug cisplatin, identifying an extensive collection of genes enriched, as expected, for components of DNA damage response pathways. More recently, Jin *et al.* [25] in the Kumar laboratory introduced transposon insertion alleles affecting over 3,600 yeast genes into a filamentous strain of yeast to screen for mutants exhibiting impaired or exaggerated filamentous growth. This analysis identified hundreds of genes contributing to filamentous growth, and, in particular, suggested a role for the mitochondrial retrograde signaling pathway in filamentous growth. Finally, Kumar *et al.* [22] used transposon-tagging to determine the subcellular localization of more than 2,500 yeast proteins during vegetative growth. In total, the transposon insertion collection underlying these studies represents the largest collection of mutant alleles constructed in any single organism to date and stands as a strong testimony to the utility of transposon technology for functional genomics.

Transposon Mutagenesis in *Arabidopsis Thaliana*

Agrobacterium-mediated T-DNA transformation is a popular tool for mutagenesis of *Arabidposis thaliana* [26]; in fact, large collections of T-DNA insertion lines have been generated and are available to the *Arabidopsis* scientific community for functional genomics studies. These mutant lines, however, often harbor complex and/or multiple insertions. As strong complimentary approaches, two transposon-based mutagenesis systems have been successfully developed in this model organism: the maize-originated Activator/Dissociation (*Ac/Ds*) and Enhancer/Suppressor-mutator (*En/Spm*) [27-28].

The *Ac/Ds* system was first developed in 1993 [27], building on the fact that the maize transposable elements *Ac/Ds* are active in a wide range of dicotyledonous and monocotyledonous species, including *Arabidopsis thaliana*. This two-component system offers several advantages over T-DNA-based approaches. First, transposon mutagenesis provides a greater degree of control regarding the mutagenesis process itself and, accordingly, a decreased risk of off-site mutations. By T-DNA-based approaches, mutations often arise after transformation; however, by transposon mutagenesis, the mutagenic event is separated from the transformation process. In this case, transposon and transposase are transformed into separated lines, and transposition only occurs after the plant line carrying the transposon is crossed with the transposase line. Second, these two-component systems provide a simple way to confirm that an observed mutant phenotype is caused by transposon insertion. The mutation caused by a transposition-defective transposon is stable as long as the plant does not express transposase; however, such a mutation can be reverted by removal of the transposon upon introduction of the transposase gene. Additionally, transposon systems can be very valuable when mutagenesis of a particular region of chromosomal DNA is desired. Most plant transposons tend to exhibit short-range transposition, called "local hopping". Therefore, a transposon insertion library will be enriched for insertions into genetically linked sites. This, of course, presents a potential downside, however, if the intended purpose is to build an insertion mutant library with maximum genome coverage.

In order to efficiently construct a collection of stable independent insertions over the whole genome, a similar transposon-based system has been developed using *En/Spm* elements [28-30]. In contrast to the *Ac/Ds* system, where selection for kanamycin resistance is used, BAR and SU1 genes serve as selectable markers in the *En/Spm* system for the selection of soil-grown plants, thereby avoiding the need to prepare sterile media and seeds. Additionally, instead of having two separated lines for transposon and transposase, the nonautonomous defective *Spm* (*dSpm*) and *Spm* transposase are contained in one single T-DNA, greatly reducing the number of progeny that need to be screened. However, the obvious drawback of having two components on one single T-DNA is the instability of transposants, which requires a large number of independent primary T-DNA transformants for starter lines. To minimize the number of starter lines needed for this approach, a modified four-element transposon system combining *Ac/Ds* and *Spm/dSpm* elements has been reported recently [31].

P Elements in *Drosophila Melanogaster*

Transposable elements have been studied extensively in *Drosophila*, in part, because transposons account for as much as 15% of the *D. melanogaster* genome. In total, approximately 30 classes of transposons have been found in this species, with P elements receiving the bulk of attention subsequent to their use for transgenesis by Rubin and Spradling [32-33]. In the years since, modified P elements have been widely used for gene tagging, gene disruption, chromosome engineering and inducible gene expression.

Most notably, the P element mutagenesis system has been used to generate a large collection of fly mutant lines for phenotype analysis [34]. The most efficient system for P element mutagenesis is binary, including an immobile transposase gene and one or more transposable, but nonautonomous P elements. These two parts of the P element are separated to prevent uncontrolled transposition. A variety of engineered P elements have been constructed, each carrying a selectable marker gene — generally eye color markers or the neomycin resistance gene — allowing easy identification of transformants. This represents the major advantage of P element mutagenesis over traditional chemical or radiation-based methods, wherein tedious positional cloning and library screens are often necessary. After microinjection into pre-blastoderm embryos and random insertion into the fly genome, the nonautonomous P element constructs are ready to be mobilized by introduction of the transposase into the embryo. Over the last ten years, numerous projects have generated collections of P-element mutations, and, thus far, P element insertions have been associated with 4,216 of 14,000 predicted genes in *D. melanogaster*. In total, 9,161 P-element alleles are listed in the *Drosophila* database Flybase [35].

In addition to phenotypic screening, P elements can also be used for the identification of enhancer elements by enhancer trapping [36-38]. To serve as an enhancer trap, the P-element constructs carry a reporter gene controlled by a weak basal promoter (Figure 3A). The *E. coli lacZ* gene is commonly used as this reporter due to the ease with which □-galactosidase can be detected. Upon P element insertion, *lacZ* expression reflects the activity of adjacent regulatory elements; accordingly, insertion of the P element trap near an enhancer results in easily detected □-galactosidase activity. Furthermore, because most enhancers direct cell- or tissue-specific gene expression, insertions near an enhancer usually lead to unique *lacZ* expression patterns within particular cells or tissues. P element enhancer traps have also been engineered to allow the cloning of genomic DNA near any particular insertion site, making it possible to isolate the enhancer and its target genes from genomic DNA. This approach has recently been extended to allow for the study of spatial gene expression patterns in real time using green fluorescent protein (GFP) or luciferase reporters [38].

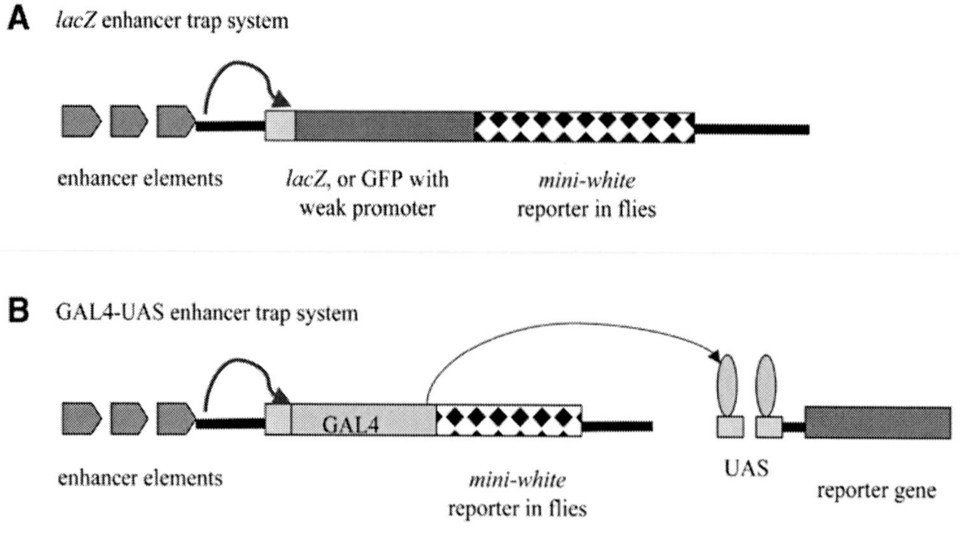

Figure 3. Transposon-based enhancer traps in *D. melanogaster*. A) Diagram of a *lacZ*-based enhancer trap wherein insertion of the transposable element near an enhancer results in expression and detection

of a reporter gene driven from a weak basal promoter. B) Schematic representation of the GAL-UAS system. Transposition near an enhancer results in *GAL4* expression, and the yeast Gal4p activator recognizes the UAS sequences upstream of a GFP reporter gene. The *mini-white* gene serves as a selectable marker in flies. Elements represented in this figure are not drawn to scale.

To date, many large libraries of *lacZ* P-element stocks have been generated, encompassing over 3,700 enhancer trap lines. These lines have found wide use in studies of enhancers and genes directing highly specific cell expression.

In spite of its success in many cell lines and tissues, applications of *lacZ*-based enhancer traps are limited in neuronal research: nuclear-targeted □-galactosidase cannot reveal the shape of neurons, making it difficult to correlate specific neuronal elements with staining patterns. Many of these limitations were addressed by the development of the GAL4-UAS enhancer trap strategy [39]. In the binary GAL4-UAS system (Figure 3B), the *lacZ* reporter is replaced with the yeast *GAL4* gene, encoding a transcriptional activator that is also functional in *Drosophila*. Rather than visualize *GAL4* expression directly, the GAL4 activator is designed to drive secondary reporters (commonly *lacZ* or GFP) controlled by a GAL4-responsive upstream activation sequence (UAS). With this system, GAL4 activity can be easily monitored by virtue of the second reporter. The GAL4-UAS system has been successfully applied in neuronal research because it allows visualization of cell bodies and, in neurons, axons and dendrites. For example, this technique has been used to observe selectively the axonal processes corresponding to intrinsic neurons of Drosophila mushroom bodies. In this study, the authors also described an unexpected degree of anatomical subcomponents within the mushroom bodies, based on discrete patterns of reporter gene expression in subsets of Kenyon cells [39].

Applications of the Tc1 Transposon Superfamily

The Tc*1*/*mariner* superfamily represents a diverse set of eukaryotic transposons defined by transposase sequence homologies and a similar cut-and-paste mechanism of transposition [40]. Transposons in this superfamily possess a common structure, with transposase coding sequence in the center of the element flanked by inverted repeats at the ends (Figure 4A). Tc*1*, the first *Caenorhabditis elegans* transposon to be identified, and insect *mariner* elements are the founding members of this superfamily. Tc*1*-like transposons are extraordinarily widespread in nature, presumably as a result of horizontal transmission of ancient transposons [41-42]. The vast majority of Tc*1*/*mariner* elements, however, are immobile in all eukaryotic genomes, and only one active element has been identified from vertebrates to date. Most of these elements have been inactivated by mutations through a process termed "vertical inactivation".

Interestingly, transposases from the Tc*1* family seem to have no requirement for highly specific host factors during transposition; recombinant transposase proteins from this family are capable of mediating complete transposition reactions *in vitro* [40]. With no requirement for host-specific factors in transposition, Tc*1* family transposons have emerged as attractive candidates for the development of vertebrate transposon-based mutagenesis systems. In fact, scientists have successfully mobilized Tc*1*/*mariner* elements by genetic engineering, and the two most powerful vertebrate transposons system, *Sleeping Beauty* (*SB*) and *Frog Prince* (*FP*), were generated by molecular reconstructions of defective fish and frog Tc*1*/*mariner* family transposon sequences, respectively [43].

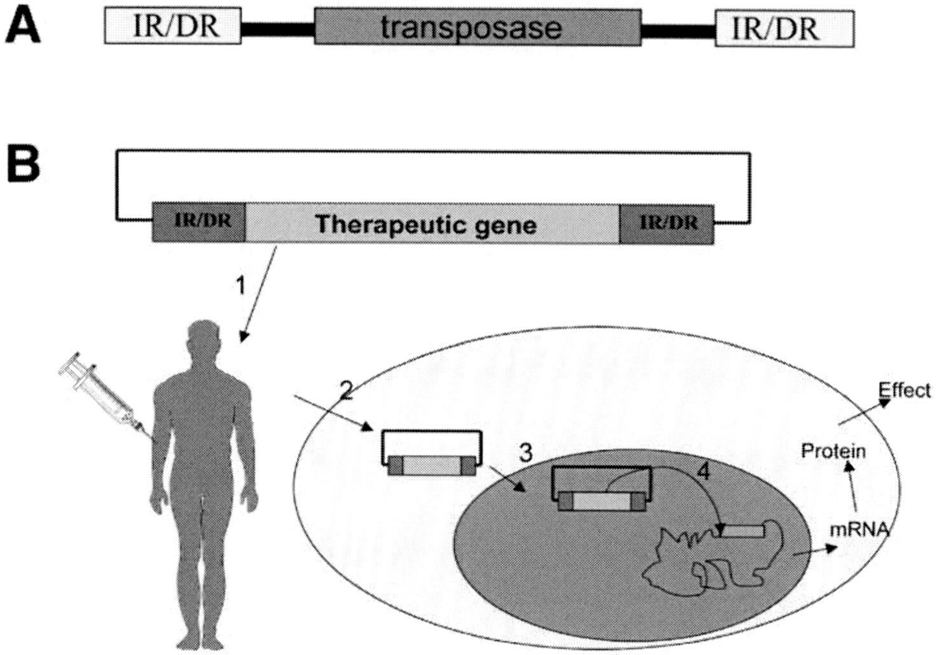

Figure 4. The *Sleeping Beauty* transposon and its potential application as a vector for gene therapy. A) Simplified structure of the Tc*1*/ *mariner* family of transposons. B) Schematic overview illustrating the potential application of *Sleeping Beauty* for non-viral gene therapy. The transposase gene has been replaced with a desired therapeutic gene. Introduction into the cell, passage across the nuclear membrane, and chromosomal integration is indicated in steps 2-4, respectively.

The Sleeping Beauty Transposon

Sleeping Beauty is a genetically reconstructed Tc*1*-like transposon from salmonid fish [43], which, at the time of its development, represented the first active DNA-based transposon system from vertebrates for genetic transformation and insertional mutagenesis. The *SB* system is binary, consisting of a functional transposase and a mobile, nonautonomous salmonid-type transposon. The synthetic *SB* transposase was genetically engineered through a 10-step process, while the transposable element, the T element, was derived from a single salmonid-type Tc*E* transposon from *Tanichthys albonubes*. The inverted repeats of this T element were genetically modified in order to improve the overall transposition efficiency of the *SB* system [44]. *SB* is highly efficient in terms of transposition: the rate of transposition mediated by *SB* is nearly an order of magnitude higher than those observed for a variety of transposons from nematodes and flies in mammalian cells [45-46]. The *SB* system has proven to be a successful and popular method for insertional mutagenesis in mice, enabling the generation of many mutant lines for subsequent phenotypic study [47-49]. Relative to other transposons, *Sleeping Beauty* offers several advantages. First, upon transposition, the *SB* element tends to reintegrate near the original donor site (DS) in the mouse germ line — a process termed “local hopping”. This local hopping may result from unequal availability of chromosomes as transposition targets in the mouse nucleus. Specifically, approximately 50% of *SB* integrations are clustered within a 3Mb region around the DS, and 80% are located on the same chromosome as the DS. This property of *Sleeping Beauty* provides a means by which insertional mutagenesis may be directed to a particular gene cluster or chromosomal region of interest. On the other hand, *SB*-based genome-wide mutagenesis is still feasible, due to the high number of transposition sites in the germ cells of founder mice and the fact that approximately 20% of the remaining transpositions are randomly distributed throughout the genome. The percentage of random transposition can be significantly improved by co-injecting SB transposons and transposase mRNA into single-cell mouse embryos; by this approach, integration into any chromosome occurs with equal likelihood.

In addition to its applications in mice, *SB*-based transgenesis has been used to establish enhancer trap lines in another vertebrate model system, the zebrafish *Danio rerio* [50-51]. Similar to enhancer trap lines in *D. melanogaster*, zebrafish lines provide high-fidelity information characterizing tissue- or organ-specific gene expression patterns by GFP reporter fluorescence. As most enhancer trap lines only harbor a single transposon insertion, the insertion site responsible for a given GFP expression pattern can be readily identified. For example, Balciunas *et al.* [52] constructed an enhancer trap line harboring a transposon insertion near the *mkp3* locus, a gene known to be regulated in a FGF8-dependent pattern. In this study, Balcuunas and colleagues found an expected GFP expression pattern in the midbrain-hindbrain boundary, forebrain, and ventricle, matching previous knowledge regarding the expression pattern of FGF8. GFP expression in another enhancer trap line, however, showed specific enrichment in caudal primary motoneurons, due to an insertion into the poly (ADP-ribose) glycohydrolase (*PARG*) locus. This surprising expression pattern was confirmed by *in situ* hybridization techniques for the endogenous *PARG* mRNA, suggesting that this enhancer trap line can be used as a high-fidelity indicator for the unexpected and highly localized expression of *PARG*.

Potential Applications of Sleeping Beauty Towards Gene Therapy

Non-viral DNA-mediated gene therapy has been considered at length as an approach to express therapeutic genes; however, the major limitation in DNA-mediated gene therapy lies in the relatively short duration of target gene expression. To overcome this limitation, scientists have explored the possibility of utilizing the *Sleeping Beauty* system as a gene delivery vehicle [53-55]. Similar to viral vectors, the *SB* system can be used to integrate therapeutic genes into the host genome, improving the stable expression of newly introduced DNA sequences by more than 100-fold (Figure 4B). In addition, *SB* vectors are not infectious and, hence, do not pose bio-safety issues. These features have prompted numerous pilot studies into the application of *SB* transposons as vectors for gene therapy. In particular, the *SB* transposon system has been used in hemophilic mice to complement both factor VIII and factor IX deficiencies, and *SB* has shown efficacy in the treatment of murine tyrosinemia type I or fumarylacetoacetate hydrolase deficiency. *SB* has also been used extensively to correct epidermal holoclones from unrelated junctional epidermolysis bullosa patients and has shown the ability to mediate insulin expression in the livers of type I diabetic mice [56-58].

An early study indicating the feasibility of *SB* as a means of mediating long-term gene expression in mouse tissues was reported by Yant *et al.* [59]. Using the SB system, Yant and colleagues successfully integrated the gene encoding human clotting factor IX into the genome of factor IX-deficient mice. By assaying for secreted human clotting factor IX in the mouse blood stream, they demonstrated extended and stable expression from the therapeutic gene, suggesting the ability of *SB* to mediate gene transfer in the liver.

SB-mediated gene transfer has also been investigated in hematopoietic stem cells (HSC), where gene therapy can potentially cure a variety of human hematopoietic diseases, such as sickle cell disease. For gene transfer in primary HSC, conventional non-viral DNA-mediated vectors have been limited to applications in myeloid-erythroid colony-forming cells by electroporation. To overcome this limitation, M. Noll *et al.* [60] have successfully utilized the *SB* system to mediate transfer of the human FANCC gene into primitive HSC in a mouse model of Fanconi anemia complementation group. In this work, the authors demonstrated selective outgrowth of HSC following the introduction of human FANCC cDNA, indicating stable FANCC expression in HSC.

Compared against conventional non-viral vectors, the *Sleeping Beauty* system offers several advantages. First, *SB* insertions are highly efficient and stable, allowing for the long lasting expression of a therapeutic gene. Second, *SB*-mediated gene transfer can be controlled easily by the supply of a second vector carrying the transposase gene; this feature helps to avoid unintentional transpositions. Third, *SB* transposons per se are not rigorously limited by the size of therapeutical genes they carry. Additionally, the *SB* transposases contain a nuclear localization signal that may enhance the efficiency of transposon delivery from the cytoplasm to the nucleus. However, this translocation process is still the major hurdle for the *SB* system, because both the cell and nuclear membranes represent barriers for DNA delivery. Thus, to date, preclinical applications of SB have been limited to diseases where these barriers can be overcome.

The Frog Prince Transposon

Although *Sleeping Beauty* shows no host-restrictions in vertebrates, it is not equally efficient in all model species [61]: transposition assays in a variety of vertebrate cultured cells have revealed extensive variability in the efficiency of *SB* transposition. Thus, the *Frog Prince* (*FP*), another highly active vertebrate transposon, was developed with the intent of extending applications of transposon-based genetic manipulations [62].

FP is a Tc*1*/*mariner*-like element that was recently reactivated from genomic transposon copies of the Northern Leopard Frog (*Rana pipiens*). With a similar structure to the *SB* transposon, the *FP* system is also binary, consisting of a genetically reconstructed transposase and transposable DNA substrate with specific inverted repeats. The DNA sequences in the *SB* and *FP* transposons are, however, clearly distinct, sharing only 50%

similarity in the transposase-coding region. The *FP* transposon is active in cell lines of all major vertebrate species and shows similar, or sometimes even higher, transposition activity, as compared to the *SB* system. For example, *FP* actually exhibits 70% higher transposition efficiency than the fish-originated *SB* system in zebrafish cells [63]. Due to its highly efficient transposition, the *FP* system has emerged as an attractive molecular tool for insertional mutagenesis in many vertebrate cells. Furthermore, the lack of detectable interference between *SB* and *FP* offers the possibility that the two systems could be used simultaneously and complementarily for genetic analyses in vertebrates.

Applications of Transposons for Forensic Genomics

Transposable elements have proven useful in a wide variety of biological disciplines, and forensic genomics stands as an interesting example. Specifically, Human *Alu* element-based genetic systems, combined with high-throughput DNA analysis, have become powerful tools in forensic science. In particular, *Alu*-based approaches have served an important role in forensic applications related to human gender identification [63-65]. In forensic studies, gender identification is typically based on sex chromosome difference at the *Amelogenin* locus [65]. By this approach, the gender of human DNA samples may be determined by considering the differently sized PCR products generated from the X and Y chromosomal alleles of the *Amelogenin* gene, named *AMELX* and *AMELY*, respectively. In some instances, however, the *AMELY* allele may be lost in males; thus, the *Amelogenin* locus does not provide a foolproof determinant of gender [66]. In complement, unique *Alu* insertions on the sex chromosomes, *AluSTXa* and *AluSTYa*, provide an effective indicator of chromsome identity [64]. In combination, these approaches provide added assurance for the accuracy of gender test results, since co-occurrence of independent mutations on two separate genomic locations is extremely unlikely. In addition, since *AluSTXa* and *AluSTYa* are fixed on the sex chromosomes, fluorescent-based detection of these markers can be used as quantitative assay for male and female contributions to a mixed DNA sample.

The identification of human DNA in a mixed sample is another common problem in forensic labs. Most human forensic casework involves the analysis of nuclear loci, which requires quantification of human nuclear DNA. As a rich source of human genetic markers, *Alu* elements have been widely used for quantifying human nuclear DNA [67-69]. For example, intra-*Alu*-based PCR assays have been developed for human-specific DNA identification and quantification based on the *Alu* subfamiliesYb8 and Yd6, taking advantage of the large numbers of unique insertion or deletion in these subfamilies [68]. The intra-Yb8-based PCR assay has a linear quantification range of 10–0.001 ng of DNA and a detection limit as low as 10 pg of human DNA from a mixed source sample. The intra-Yd6-based PCR assay is more specific to human DNA; it often helps to avoid background amplification from non-human samples, although the assay is less sensitive quantitatively with a detection limit of 100 pg of human DNA. The combination of these two assays in forensic genomics can provide a powerful tool for sensitive human DNA identification and quantification from mixed sources.

Conclusion

As summarized in this chapter, transposons have emerged as informative and widely varied tools for application in a diverse spectrum of biological disciplines. Most notably, transposon-based systems have been utilized on a genome-wide scale for insertional mutagenesis and gene/enhancer trapping, with application of the Tn*10*-derived T-POP transposon for microbial genomics, bacterial and yeast transposon derivatives for mutagenesis of *S. cerevisiae*, *Ac/Ds* and *En/Spm* elements in *A. thaliana*, and P elements in *Drosphila melanogaster*. Thus, in prokaryotes and eukaryotes alike, modified transposable elements have served as workhorses for the generation of large collections of gene disruption mutants. Interestingly, however, in many prokaryotic and lower eukaryotic organisms, advances in large-scale cloning technologies have somewhat obviated the need for transposon-based approaches. Many of these approaches are not as easily applied in higher eukaryotes; accordingly, transposon technology remains very relevant in such organisms. Furthermore, transposon-based traps still constitute unique and extremely valuable biological tools, providing one of the best mechanisms for gene identification and discovery.

Transposon technology promises to play an increasing role as a tool for the study of vertebrate biology. Towards this end, Tc*1*/*mariner* transposons have been developed into a technology platform for vertebrate genetics with applications encompassing gene therapy, transgenesis, somatic mutagenesis and germ line mutagenesis for gene discovery. In particular, synthetic *Sleeping Beauty* and *Frog Prince* transposon systems have been widely used for forward genetics in many model vertebrate organism including mice and zebrafish. *SB* has also become an attractive vector for non-viral gene therapy. Recent studies in mice and HSC have suggested the potential in *SB* as a vector for the stable, long-lasting expression of therapeutic genes. In addition, human *Alu* elements have found important applications in many areas of forensic science. Collectively, transposons represent a malleable platform for a broad range of studies, with continuing, if not expanding, utility in genomic studies of higher eukaryotes and vertebrates.

References

[1] Berg, D. E., Berg, C. M. and Sasakawa, C. (1984). Bacterial transposon tn5: Evolutionary inferences. *Mol Biol Evol.*, *1*, 411-422.

[2] Craig, N. L. (1991). Tn7: A target site-specific transposon. *Mol. Microbiol.*, *5*, 2569-2573.

[3] Deininger, P. L. and Batzer, M. A. (1999). Alu repeats and human disease. *Mol. Genet. Metab.*, *67*, 183-193.

[4] Kumar, A., Vidan, S. and Snyder, M. (2002). Insertional mutagenesis: Transposon-insertion libraries as mutagens in yeast. *Methods Enzymol*, *350*, 219-229.

[5] Winterberg, K. M. and Reznikoff, W. S. (2007). Screening transposon mutant libraries using full-genome oligonucleotide microarrays. *Methods Enzymol*, *421*, 110-125.

[6] Wheelan, S. J., Scheifele, L. Z., Martinez-Murillo, F., Irizarry, R. A. and Boeke, J. D. (2006). Transposon insertion site profiling chip (tip-chip). *Proc Natl Acad Sci U S A, 103*, 17632-17637.

[7] Seringhaus, M., Kumar, A., Hartigan, J., Snyder, M. and Gerstein, M. (2006). Genomic analysis of insertion behavior and target specificity of mini-tn7 and tn3 transposons in saccharomyces cerevisiae. *Nucleic Acids Res, 34*, e57.

[8] Hensel, M., Shea, J. E., Gleeson, C., Jones, M. D., Dalton, E. and Holden, D. W. (1995). Simultaneous identification of bacterial virulence genes by negative selection. *Science, 269*, 400-403.

[9] Rosas-Magallanes, V., Stadthagen-Gomez, G., Rauzier, J., Barreiro, L. B., Tailleuz, L., Boudou, F., Griffin, R., Nigou, J., Jackson, M., Gicquel, B. and Neyrolles, O. (2007). Signature-tagged transposon mutagenesis identifies novel mycobacterium tuberculosis genes involved in the parasitism of human macrophages. *Infect. Immun., 75*, 504-507.

[10] Camacho, L. R., Constant, P., Raynaud, C., laneelle, M. A., Triccas, J. A., Gicquel, B., Daffe, M. and Guilhot, C. (2001). Analysis of the phthiocerol dimycocerosate locus of mycobacterium tuberculosis. *J. Biol. Chem., 276*, 19845-19854.

[11] Rappleye, C. A. and Roth, J. R. (1997). A tn10 derivative (t-pop) for isolation of insertions with conditional (tetracycline-dependent) phenotypes. *J. Bacteriol, 179*, 5827-5834.

[12] Hidalgo, A. A., Trombert, A. N., Castro-Alonso, J. C., Santiviago, C. A., Tesser, B. R., Youderian, P. and Mora, G. C. (2004). Insertions of mini-tn10 transposon t-pop in salmonella enterica sv. Typhi. *Genetics, 167*, 1069-1077.

[13] Way, J. C., Davis, M. A., Morisato, D., Roberts, D. E. and Kleckner, N. (1984). New tn10 derivatives for transposon mutagenesis and for construction of lacz operon fusions by transposition. *Gene, 32*, 369-379.

[14] Judson, N. and Mekalanos, J. (2000). Tnaraout, a transposon-based approach to identify and characterize essential bacterial genes. *Nat. Biotechnol., 18*, 740-745.

[15] Jacobs, M. A., Alwood, A., Thaipisuttikul, I., Spencer, D., Haugen, E., Ernst, S., Will, O., Kaul, R., Raymond, C., Levy, R., Chun-Rong, L., Guenthner, D., Bovee, D., Olson, M. V. and Manoil, C. (2003). Comprehensive transposon mutant library of pseudomonas aeruginosa. *Proc Natl Acad Sci U S A, 100*, 14339-14344.

[16] Smith, V., Botstein, D. and Brown, P. O. (1995). Genetic footprinting: A genomic strategy for determining a gene's function given its sequence. *Proc Natl Acad Sci U S A, 92*, 6479-6483.

[17] Ji, H., Moore, D. P., Blomberg, M. A., Braiterman, L. T., Voytas, D. F., Natsoulis, G. and Boeke, J. D. (1993). Hotspots for unselected ty1 transposition events on yeast chromosome iii are near trna genes and ltr sequences. *Cell, 73*, 1007-1018.

[18] Ross-Macdonald, P., Sheehan, A., Roeder, G. S. and Snyder, M. (1997). A multipurpose transposon system for analyzing protein production, localization, and function in *saccharomyces cerevisiae. Proc. Natl. Acad. Sci. USA, 94*, 190-195.

[19] Ross-Macdonald, P., Coelho, P. S., Roemer, T., Agarwal, S., Kumar, A., Jansen, R., Cheung, K. H., Sheehan, A., Symoniatis, D., Umansky, L., Heidtman, M., Nelson, F. K., Iwasaki, H., Hager, K., Gerstein, M., Miller, P., Roeder, G. S. and Snyder, M.

(1999). Large-scale analysis of the yeast genome by transposon tagging and gene disruption. *Nature*, *402*, 413-418.

[20] Kumar, A., DesEtages, S., Coelho, P., Roeder, G. and Snyder, M. (2000). High-throughput methods for the large-scale analysis of gene function by transposon tagging. *Methods Enzymol*, *328*, 550-574.

[21] Kumar, A., Seringhaus, M., Biery, M., Sarnovsky, R. J., Umansky, L., Piccirillo, S., Heidtman, M., Cheung, K.-H., Dobry, C. J., Gerstein, M., Craig, N. and Snyder, M. (2004). Large-scale mutagenesis of the yeast genome using a tn7-derived multipurpose transposon. *Genome Research*, *14*, 1975-1986.

[22] Kumar, A., Agarwal, S., Heyman, J. A., Matson, S., Heidtman, M., Piccirillo, S., Umansky, L., Drawid, A., Jansen, R., Liu, Y., Cheung, K. H., Miller, P., Gerstein, M., Roeder, G. S. and Snyder, M. (2002). Subcellular localization of the yeast proteome. *Genes Dev*, *16*, 707-719.

[23] Kumar, A., Harrison, P. M., Cheung, K. H., Lan, N., Echols, N., Bertone, P., Miller, P., Gerstein, M. B. and Snyder, M. (2002). An integrated approach for finding overlooked genes in yeast. *Nat Biotechnol*, *20*, 58-63.

[24] Bachman, N., Biery, M., Boeke, J. and Craig, N. (2002). Tn7-mediated mutagenesis of saccharomyces cerevisiae genomic DNA in vitro. *Methods Enzymol*, *350*, 230-247.

[25] Jin, R., Dobry, C. J., McCown, P. J. and Kumar, A. (2007). Large-scale analysis of yeast filamentous growth by systematic gene disruption and overexpression. *Mol Biol Cell*,

[26] Krysan, P. J., Young, J. C. and Sussman, M. R. (1999). T-DNA as an insertional mutagen in *arabidopsis*. *Plant Cell 1999*, *11*, 2283-2290.

[27] Long, D., Martin, M., Sundberg, E., Swinburne, J., Puangsomlee, P. and Coupland, G. (1993). The maize transposable element system ac/ds as a mutagen in arabidopsis: Identification of an albino mutation induced by ds insertion. *Proc Natl Acad Sci U S A*, *90*, 10370-10374.

[28] Speulman, E., Metz, P. L., Arkel, G. v., Hekkert, B. e. L., Stiekema, W. J. and Pereira, A. (1999). A two-component enhancer-inhibitor transposon mutagenesis system for functional analysis of the *arabidopsis* genome. *Plant Cell*, *11*, 1853-1866.

[29] Machida, C., Onouchi, H., Koizumi, J., Hamada, S., Semiarti, E., Torikai, S. and Machida, Y. (1997). Characterization of the transposition pattern of the ac element in arabidopsis thaliana using endonuclease i-scei. *Proc Natl Acad Sci U S A*, *94*, 8675-8680.

[30] Tissier, A. F., Marillonnet, S., Klimyuk, V., Patel, K., Torres, M. A., Murphy, G. and Jones, J. D. (1999). Multiple independent defective suppressor-mutator transposon insertions in *arabidopsis*: A tool for functional genomics. *11*, 1841-1852.

[31] Panjabi, P., Burma, P. K. and Pental, D. (2006). Use of the transposable element ac/ds in conjunction with spm/dspm for gene tagging allows extensive genome coverage with a limited number of starter lines: Functional analysis of a four-element system in arabidopsis thaliana. *Mol. Genet. Genomics*, *276*, 533-543.

[32] Karess, R. E. and Rubin, G. M. (1984). Analysis of p transposable element functions in drosophila. *Cell*, *38*, 135-146.

[33] Spradling, A. C., Stern, D. M., Kiss, I., Roote, J., Laverty, T. and Rubin, G. M. (1995). Gene disruptions using *p* transposable elements: An integral component of the drosophila genome project. *Proc. Natl. Acad. Sci. USA*, *92*, 10824-10830.

[34] Spradling, A. C., Stern, D., Beaton, A., Rhem, E. J., Laverty, T., Mozden, N., Misra, S. and Rubin, G. M. (1999). The berkeley drosophila genome project gene disruption project: Single p-element insertions mutating 25% of vital drosophila genes. *Genetics*, *153*, 135-177.

[35] Consortium, F. (2003). The flybase database of the drosophila genome projects and community literature. *Nucleic Acids Res.*, *31*, 172-175.

[36] Pignoni, F., Hu, B. and Zipursky, S. L. (1997). Identification of genes required for drosophila eye development using a phenotypic enhancer-trap. *Proc Natl Acad Sci U S A*, *94*, 9220-9225.

[37] Wilson, C., Pearson, R. K., Bellen, H. J., O'Kane, C. J., Grossniklaus, U. and Gehring, W. J. (1989). P-element-mediated enhancer detection: An efficient method for isolating and characterizing developmentally regulated genes in drosophila. *Genes Dev*, *3*, 1301-1313.

[38] Morin, X., Daneman, R., Zavortink, M. and Chia, W. (2001). A protein trap strategy to detect gfp-tagged proteins expressed from their endogenous loci in drosophila. *Proc. Natl. Acad. Sci. USA*, *98*, 15050-15055.

[39] Sepp, K. J. and Auld, V. J. (1999). Conversion of lacz enhancer trap lines to gal4 lines using targeted transposition in drosophila melanogaster. *Genetics*, *151*, 1093-1101.

[40] Robertson, H. M. (1995). The tc1-mariner superfamily of transposons in animals. *J. Insect Physiol.*, *41*, 99-105.

[41] Lohe, A. R., Moriyama, E. N., Lidholm, D. A. and Hartl, D. L. (1995). Horizontal transmission, vertical inactivation, and stochastic loss of mariner-like transposable elements. *Mol. Biol. Evol.*, *12*, 62-72.

[42] Kidwell, M. G. (1992). Horizontal transfer. *Curr. Opin. Genet. Dev.*, *2*, 868-873.

[43] Ivics, Z., Hackett, P. B., Plasterk, R. H. and Izsvak, Z. (1997). Molecular reconstruction of sleeping beauty, a tc1-like transposon from fish, and its transposition in human cells. *Cell*, *91*, 501-510.

[44] Zayed, H., Izsvak, Z., Walisko, O. and Ivics, Z. (2004). Development of hyperactive sleeping beauty transposon vectors by mutational analysis. *Mol. Ther.*, *9*, 292-304.

[45] Izsvák, Z., Ivics, Z. and Plasterk, R. H. (2000). Sleeping beauty, a wide host-range transposon vector for genetic transformation in vertebrates. *J. Mol. Biol.*, *302*, 93-102.

[46] Liu, G., Geurts, A. M., Yae, K., Srinivasan, A. R., Fahrenkrug, S. C., Largaespada, D. A., Takeda, J., Horie, K., Olson, W. K. and Hackett, P. B. (2005). Target-site preferences of sleeping beauty transposons. *J. Mol. Biol.*, *346*, 161-173.

[47] Horie, K., Kuroiwa, A., Ikawa, M., Okabe, M., Kondoh, G., Matsuda, Y. and Takeda, J. (2001). Efficient chromosomal transposition of a tc1/mariner- like transposon sleeping beauty in mice. *Proc Natl Acad Sci U S A*, *98*, 9191-9196.

[48] Dupuy, A. J., Fritz, S. and Largaespada, D. A. (2001). Transposition and gene disruption in the male germline of the mouse. *Genesis*, *30*, 82-88.

[49] Carlson, C. M., Dupuy, A. J., Fritz, S., Roberg-Perez, K. J., Fletcher, C. F. and Largaespada, D. A. (2003). Transposon mutagenesis of the mouse germline. *Genetics*, *165*, 243-256.

[50] Davidson, A. E., Balciunas, D., Mohn, D., Shaffer, J., Hermanson, S., Sivasubbu, S., Cliff, M. P., Hackett, P. and Ekker, S. (2003). Efficient gene delivery and gene expression in zebrafish using sleeping beauty transposons. *Dev. Biol.*, *263*, 191-202.

[51] Amsterdam, A., Burgess, S., Golling, G., Chen, W., Sun, Z., Townsend, K., Farrington, S., Haldi, M. and Hopkins, N. (1999). A large-scale insertional mutagenesis screen in zebrafish. *Genes Dev.*, *13*, 2713-2724.

[52] Balciunas, D., Davidson, A. E., Sivasubbu, S., Hermanson, S. B., Welle, Z. and Ekker, S. C. (2004). Enhancer trapping in zebrafish using the sleeping beauty transposon. *BMC Genomics*, *5*, 62.

[53] Izsvák, Z. and Ivics, Z. (2004). Sleeping beauty transposition: Biology and applications for molecular therapy. *Mol. Ther.*, *9*, 147-156.

[54] Liu, G., Aronovich, E. L., Cui, Z., Whitley, C. B. and Hackett, P. B. (2004). Excision of sleeping beauty transposons: Parameters and applications to gene therapy. *J. Gene Med.*, *6*, 574-583.

[55] Geurts, A. M., Yang, Y., Clark, K. J., Liu, G., Cui, Z., Dupuy, A. J., Bell, J. B., Largaespada, D. A. and Hackett, P. B. (2003). Gene transfer into genomes of human cells by the sleeping beauty transposon system. *Mol. Ther.*, *8*, 108-117.

[56] Montini, E. P., Held, P. K., Noll, M., Morcinek, N., Al-Dhalimy, M., Finegold, M., Yant, S. R., Kay, M. A. and Grompe, M. (2002). In vivo correction of murine tyrosinemia type i by DNA-mediated transposition. *Mol. Therap.*, *6*, 759-769.

[57] He, C. X., Shi, D., Wu, W. J., Ding, Y. F., Feng, D. M., Lu, B., Chen, H. M., Yao, J. H., Shen, Q., Lu, D. R. and Xue, J. L. (2004). Insulin expression in livers of diabetic mice mediated by hydrodynamics-based administration. *World J. Gastroenterol.*, *10*, 567-572.

[58] Ohlfest, J. R., Frandsen, J. L., Fritz, S., Lobitz, P. D., Perkinson, S. G., Clark, K. J., Key, N. S., McIvor, R. S., Hackett, P. B. and Largaespada, D. A. (2004). Phenotypic correction and long-term factor viii expression in hemophilia-a mice by immunotolerization and nonviral gene transfer using the sleeping beauty transposon system. *Blood*, *105*, 2691-2698.

[59] Yant, S. R., Meuse, L., Chiu, W., Ivics, Z., Izsvak, Z. and Kay, M. A. (2000). Somatic integration and long-term transgene expression in normal and haemophilic mice using a DNA transposon system. *Nature Genetics*, *25*, 35-41.

[60] Noll, M., Bennett, R., Yant, S., Kay, M. A. and Grompe, M. (2005). Stable correction of hematopoietic stem cells with non-viral gene transfer. *Molecular Therapy*, *11*, S404.

[61] Vigdal, T. J., Kaufman, C. D., Izsvák, Z., Voytas, D. F. and Ivics, Z. (2002). Common physical properties of DNA affecting target site selection of sleeping beauty and other tc1/mariner transposable elements. *J. Mol. Biol.* , *323*, 441-452.

[62] Miskey, C., Izsvak, Z., Plasterk, R. H. and Ivics, Z. (2003). The frog prince: A reconstructed transposon from rana pipiens with high transpositional activity in vertebrate cells. *Nucleic Acids Res*, *31*, 6873-6881.

[63] Callinan, P. A., Hedges, D. J., Salem, A. H., Xing, J., Walker, J. A., Garber, R. K., Watkins, W. S., Bamshad, M. J., Jorde, L. B. and Batzer, M. A. (2003). Comprehensive analysis of alu-associated diversity on the human sex chromosomes. *Gene*, *317*, 103-110.

[64] Hedges, D. J., Walker, J. A., Callinan, P. A., Shewale, J. G., Sinha, S. K. and Batzer, M. A. (2003). Mobile element-based assay for human gender determination. *Anal. Biochem.*, *312*, 77-79.

[65] Sullivan, K. M., Mannucci, A., Kimpton, C. P. and Gill, P. (1993). A rapid and quantitative DNA sex test: Fluorescence-based pcr analysis of x-y homologous gene amelogenin. *Biotechniques*, *15*, 636-638.

[66] Santos, F. R., Pandya, A. and Tyler-Smith, C. (1998). Reliability of DNA based sex tests. *Nat Genet.*, *18*, 103.

[67] Walker, J. A., Kilroy, G. E., Xing, J., Shewale, J., Sinha, S. K. and Batzer, M. A. (2003). Human DNA quantitation using alu element-based polymerase chain reaction. *Anal. Biochem.*, *315*, 122-128.

[68] Sifis, M. E., Both, K. and Burgoyne, L. A. (2002). A more sensitive method for the quantitation of genomic DNA by alu amplification. *J. Forensic Sci.*, *47*, 589-592.

[69] Nicklas, J. A. and Buel, E. (2003). Development of an alu-based, real-time pcr method for quantitation of human DNA in forensic samples. *J. Forensic Sci.*, *48*, 936-944.

In: DNA Transposable Elements Research
Editors: K. Yoshida, M. Aoki, pp. 23-67
ISBN: 978-1-60456-531-7
© 2008 Nova Science Publishers, Inc.

Chapter II

Bacterial Class II Catabolic Transposons

Masahiro Sota[1], Hirokazu Yano[2] and Masataka Tsuda[2]
[1]Corresponding author. Department of Biological Sciences, University of Idaho, 252 Life Science South, Moscow, ID 83844-3051, USA
[2]Department of Environmental Life Sciences, Graduate School of Life Sciences, Tohoku University, 2-1-1 Katahira, Sendai 980-8577, Japan

Abstract

Bacterial transposons encoding a variety of phenotypic traits are widely disseminated in diverse species, thus allowing the host strains to expand their niches. Catabolic transposons responsible for the degradation of man-made and/or natural organic compounds provide the host cells to use these compounds as the sources of carbon, energy, and/or nitrogen, and have been identified worldwide in various bacterial strains from the environments contaminated with herbicides, pesticides, petroleum, etc. Studies on these transposons have revealed that the catabolic genes for the compounds are often located on the class II (Tn*3*-like) transposons, many of which are known to carry resistance genes for antibiotics or heavy metals. In contrast to the insertion sequence (IS)-composite transposons, the class II transposons are able to transpose at high frequencies even when they are large (>50 kb) in sizes, allowing the wide dissemination of large gene clusters for the degradation of aromatic compounds such as toluene/xylenes, naphthalene, and carbazole. In the last two decades, we have identified many class II catabolic transposons and clarified their structures, molecular functions, and in addition the entire sequences of large self-transmissible plasmids carrying the transposons. In this review, we summarize the recent information on bacterial class II catabolic transposons, including their transposition mechanisms. Also described are the roles of these elements in establishment and evolution of the large catabolic plasmids.

Introduction

Various chemical compounds artificially created by humans (=xenobiotics) have rendered our living circumstances more convenient day by day. On the other hand, the earth has also been rife with xenobiotics such as herbicides and pesticides, as well as with naturally occurring chemicals such as petroleum released intentionally or inadvertently, due to human activities. Examples of these compounds are mono- and/or hetero-aromatic compounds, polycyclic aromatic hydrocarbons (PAHs), polychlorinated biphenyls (PCBs), dioxins, 2,4-dichloroacetic acid (2,4-D), and dichloro-diphenyl-trichloroethane (DDT). Most of these compounds are recalcitrant and persistent in nature because of their intrinsic stable chemical characteristics. Furthermore, they are often toxic, mutagenic, and/or carcinogenic to living creatures, including humans. Therefore, persistence of the chemicals in the environment is a big health concern worldwide, and their removal to mitigate the risks is an indispensable theme to preserve the ecosystems.

In the past several decades, many bacterial strains that can degrade and/or detoxify a wide variety of such harmful organic chemicals have been reported [94, 95, 98, 151, 153]. These organisms utilize them as sources of carbon, nitrogen, and/or energy. In natural environments, some compounds are completely mineralized by one strain, and others are decomposed cooperatively by bacterial communities. Genetic and biochemical information on bacterial dissimilation of such compounds has been accumulated by many studies all over the world, and the data have illustrated that the genes derived apparently from the same origin are distributed among bacteria over far phylogenetic distances. The phenomenon, so-called the horizontal gene transfer (HGT) that sometimes engenders genetic exchange across kingdoms of life, is now widely accepted as a major driving force of bacterial genomic evolution and adaptation to the changing environments [30, 42, 46, 47, 53, 65, 134, 135, 139, 151]. Transposons and plasmids, the main members of so-called mobile genetic elements (MGEs), are known as the major agents to frequently mediate HGT. Many transposons usually move between DNA molecules by transposition, and plasmids explore between cells through conjugation or transformation. The intracellular and intercellular DNA exchanges by transposons and plasmids, respectively, and their iterations over time have contributed to the spread of present catabolic genes for xenobiotics and to the subsequent rearrangements of these genes to generate new degradation pathways. Therefore, these two elements are considered key factors for the rapid evolution of bacterial catabolic activities.

After the discovery of catabolic plasmids such as SAL (for salicylate), TOL (for toluene), and CAM (for campher) in the 1970s, there has since been a significant increase of reports about catabolic plasmids in bacteria [94, 98, 143]. Genetic analyses of these plasmids have revealed several common features. Firstly, several sequential enzymatic reactions are necessary for degradation of an organic pollutant. Such cases are especially typical in the degradation of aromatic compounds. Secondly, the genes for these catabolic cascades are, in most cases, organized as one or a few operons ranging to more than 10 kb. Thirdly, most of these plasmids have self-transmissibility, which requires more than fifteen genes for the cell-to-cell contact and DNA transfer [168]. And finally, the catabolic genes are mostly carried on transposons or their remnants. Because of these features, catabolic plasmids are usually large (>50 kb) in size. Interestingly, catabolic genes are located on only limited groups of plasmids: in Gram-negative (G⁻) bacteria, incompatibility group P-1 (IncP-1), IncP-2, IncP-7, and IncP-9 plasmids from the γ-Proteobacteria, and unclassified plasmids from the α-Proteobacteria (mainly sphingomonads) and *Flavobacterium* strains [94, 98, 143]; and in Gram-positive (G⁺) bacteria, plasmids mainly found in *Arthrobacter* and *Rhodococcus* [98]. More curious, catabolic genes for naturally occurring compounds such as naphthalene and toluene are carried on IncP-2, IncP-7, and IncP-9 plasmids, while those for xenobiotics are found on IncP-1 and α-proteobacterial plasmids [83, 94, 98, 127, 143]. To better understand bacterial catabolic activities, we have investigated catabolic plasmids mainly belonging to the IncP-1, IncP-7, and IncP-9 groups, identified many transposons, and characterized their structures and molecular functions. The results have clarified that (i) many large (>15 kb) catabolic gene clusters are carried on class II (Tn*3*-like) transposons or associated with the unique class II insertion sequence IS*1071* (Table 1) [36, 87, 123, 131, 151] and (ii) specific insertion events of these transposons into plasmids lead to generation of catabolic plasmids [21, 94, 98]. The focus of this review is (i) the recent information on bacterial catabolic genes associated with class II transposons and IS*1071,* and (ii) the roles of these transposons in establishment and evolution of the large catabolic plasmids.

Table 1. Class II and IS*1071*-composite catabolic transposons

Name*	Size (kb)	Location§	Source bacterium	Degradable compounds (catabolic genes)†	Reference
Class II					
Tn*4651*	56	pWW0	*Pseudomonas putida* mt-2	Toluene, xylenes (*xyl*)	[149]
Tn*4653*	71	pWW0	*Pseudomonas putida* mt-2	Toluene, xylenes (*xyl*)	[147-149]
Tn*4655*	39	NAH7	*P. putida* G7	Naphthalene, salycilate (*nah*)	[132, 150]
Tn*4656*	37	pWW53	*P. putida* MT53	Toluene, xylenes (*xyl*)	[165]
Tn*4657*	86	pWW53	*P. putida* MT53	Toluene, xylenes (*xyl*)	[165]
Tn*4660*	62	pWW53	*P. putida* MT53	Toluene, xylenes (*xyl*)	[165]
Tn*4676*	73	pCAR1	*P. resinovorans* CA10	Carbazole (*car, ant*)	[73, 125]
Tn*Had2*	16	pUO1	*Delftia acidovorans* B	Haloacetates (*dehH1, dehH2*)	[129, 130]
IS*1071*-composite					
Tn*5271*	17	pBRC60	*Comamonas testosteroni* BR60	3-Chlorobenzoate (*cbaABC*)	[87]
Tn*5530*	41	pIJB1	*Burkholderia cepacia* 2a	2,4-D (*tfd, mdc*)	[105]
Tn*CNB1*	45	pCNB1	*Comamonas sp.* CNB-1	4-Chloronitrobenzene (*cnb*)	[72]
Tn*Had1*	8.9	pUO1	*D. acidovorans* B	Haloacetate (*dehH2*)	[129, 130]
Tn*Mnb1*	17	Chr	*Comamonas* sp. JS46	3-Nitrobenzoate (*mnb*)	[108]

Table 1. (Continued)

Name*	Size (kb)	Location§	Source bacterium	Degradable compounds (catabolic genes)†	Reference
N/A	42	pEST4011	*Achromobacter xylosoxidans* EST4002	2,4-D (*tfd, mdc*)	[156]
N/A	42	pJP4	*Cupriavidus eutrohpa* JMP134	2,4-D (*tfd*)	[146]
N/A	52	pADP-1	*Pseudomonas sp.* ADP	Atrazine (*atzABC*)	[77]
N/A	30	Chr	*D. acidovorans* P4a	2,4-D (*tfd*)	[43]
N/A	25	Chr	*D. tsuruhatensis* AD9	Aniline (*tad*)	[71]

* The fragment flanked by intact or turncated IS*1071* sequences are defined as an IS*1071*-composite transposon. N/A, not assigned.

§ Chr, chromosome.

† Not all genes are shown: e.g., *tfd* in pEST4011 indicates *tfdABCDEKR*. See references for details.

Structural and Functional Properties of Class II Transposons

The bacterial class II transposons, also known as Tn*3*-like transposons, are basically composed of three parts: a transposase gene (*tnpA*), a resolvase gene (*tnpR*), and terminal inverted repeats (IRs) (Figure 1) [36, 123]. The *tnpA* genes are constantly 3 kb in sizes, and the *tnpR* genes usually 0.6 kb. The length of IRs varies from 38 to 110 bp. Most of them additionally encode various phenotypic traits such as antibiotic resistance, heavy metal resistance, and xenobiotic degradation, which have been incorporated during the evolutionary histories of the transposons (Figure 1A). The structure of class II transposons contrasts with the quite simple one of insertion sequences (ISs) in that most ISs carry only one gene for their own transposition (transposase gene, *tnp*) between IRs (Figure 1A). Transposition of class II transposons is usually accompanied by duplication of the transposon, which is called the "replicative transposition" or "copy-and-paste" mechanism (Figure 2A). In the first step, the TnpA protein binds to the IRs and mediates joining of donor and target molecules, which is called a "Shapiro intermediate" [122, 123]. Replication of the transposon in the intermediate gives a cointegrate of the donor and target molecules connected by two directly repeated copies of the transposon, one at each junction. In the second step, the cointegrate resolves site-specifically at the resolution (*res*) sites by the *tnpR* product (resolvase) [36, 123]. A 4- to 6-bp duplication of the target sequence in the target molecule is generated upon transposition [36, 123, 125, 151, 165]. This transposition mechanism is quite different from that of ISs that usually transpose by the mechanism called "simple insertion", "conservative transposition", or "cut-and-paste", whereby an IS element is physically separated from its donor molecule (cut), then inserted into the target molecule (paste) (Figure 2B). In the cut-and-paste transposition, no duplication of the element occurs. The transposases of the class II transposons are able to efficiently catalyze their transposition even when the *tnpA* gene and its cognate IRs are located on separate molecules, while the transposases of IS elements predominantly act on IRs in *cis* [12, 35]. This *trans*-acting feature of the class II transposases is considered to allow efficient transposition of class II transposons with very large sizes (>50 kb) (see below).

Most of the enzymes for the resolution events encoded by the class II transposons are the site-specific serine recombinases, which catalyze only the intramolecular recombination between two copies of the specific *res* sites on the same replicon [128]. In several class II transposons, the site-specific tyrosine recombinases catalyze the resolution: the TnpS protein from the toluene-catabolic transposon Tn*4651* and the TnpI proteins from the naphthalene-catabolic transposon Tn*4655* and the cryptic transposon Tn*4430* [29, 74, 132]. These tyrosine recombinases indeed catalyze both the intra- and intermolecular recombination between two attachment (*att*) sites (Figure 2C), and are alternatively designated integrases in some cases on the basis of the integrative activities [32, 90]. Such integrases are also known to play important roles in the movement of various MGEs such as bacteriophages, conjugative transposons, and genomic islands [10, 78, 90]. The phylogenetic tree of transposases does not correspond with that of the resolution proteins (serine and tyrosine recombinases), clearly indicating that the resolution systems have been incorporated into the class II transposons from other sources [36]. At present, it is not known how class II transposons acquired the resolution systems.

At present, class II transposons are categorized into four major groups (Tn*3*-, Tn*21*-, and Tn*4651*-groups and IS*1071*) on the basis of their structural organization and functional exchangeability (Figure 1) [36, 37, 131, 149]. IS*1071* is the only one class II transposon without a resolution function (see below). The Tn*21*-group transposons can be further divided into two subgroups, Tn*21*- and Tn*1722*-subgroup transposons, according to the non-interchangeability of TnpA proteins [37]. Tn*21*-group transposons are assumed to be predominant in natural bacterial communities [16]. A novel fifth group, Tn*5393*-group, can be added to class II transposons, because (i) Tn*5393* and its derivatives code for transposases with at most 40% amino-acid identity to those of the other transposons and (ii) their IRs have 81-bp sequences, which considerably differ from those in the four other groups (Figure 1B) [14, 66, 76, 136]. Most transposons and groups described above have been identified in G^- bacterial strains, but class II transposons have also been found in G^+ bacterial strains: the cryptic transposon Tn*4430* from *Bacillus thuringiensis* [74] and the vancomycin-resistance transposon Tn*1546* from *Enterococcus feacium* [2]. Therefore, new classification of class II transposons including those of G^+ origins should be accomplished.

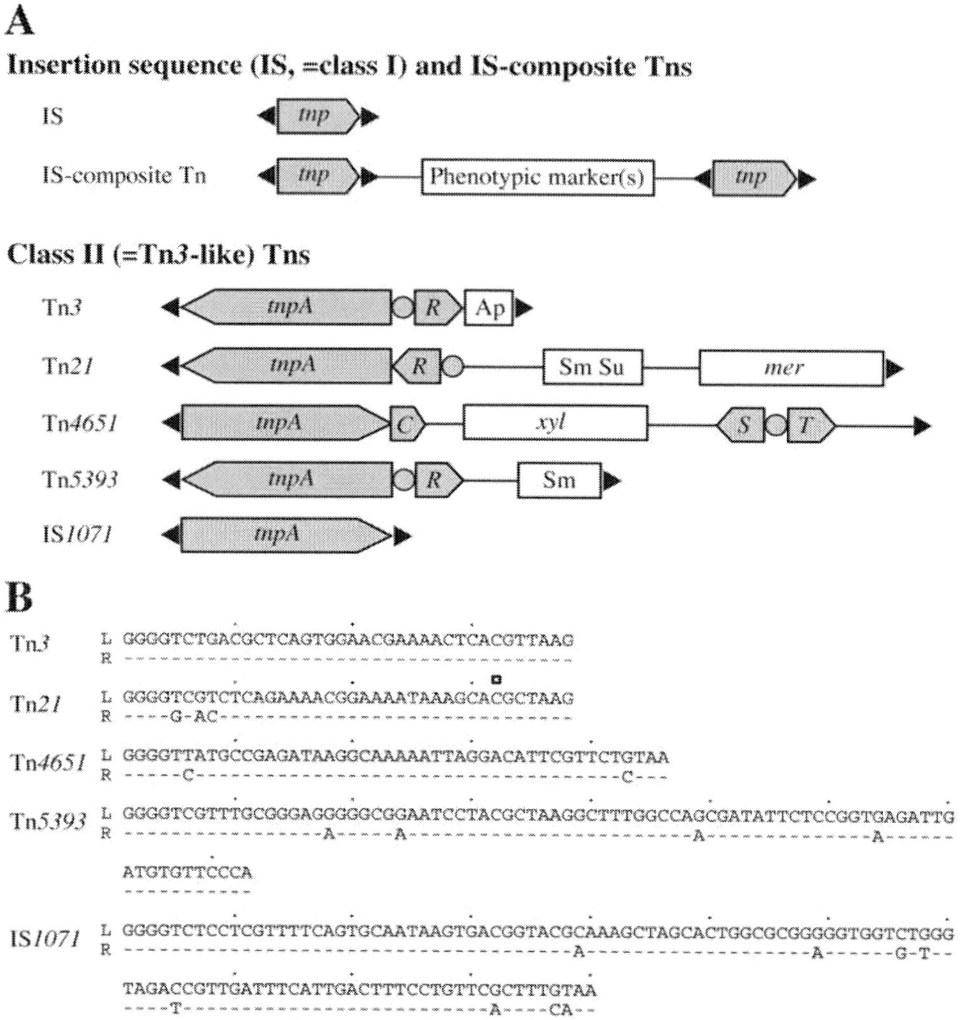

Figure 1. Schematic structures of class I and class II transposons (A) and IR sequences of class II transposons (B). (A) Class I transposons are also known as IS elements, and the fragment sandwiched by two completely or nearly identical copies of IS elements is designated IS-composite transposons. Class II transposons usually carry a *tnpA* gene, a resolution system (*tnpR-res*, *tnpI-att*, *tnpS-res-tnpT*) between IRs. IS*1071* is the only class II transposon without any resolution system. Abbreviations: *tnp*, transposase gene of an IS element; *tnpA*, transposase gene of a class II transposon; C, *tnpC* gene; R, *tnpR* resolvase gene; S, *tnpS* integrase gene; and T, *tnpT* gene; Ap, Sm, and Su, resistance to ampicillin, streptomycin, and sulfonamide, respectively; *mer*, genes for resistance to mercury; and *xyl*, genes for degradation of toluene/xylenes. The sizes of transposons and genes are arbitrary. (B) IR sequences of class II transposons in five major groups. L and R indicate the left and right IR sequences, respectively, corresponding to the figure depicted in the panel A. Dashes represent the same nucleotides as those in the left IR. Periods are placed every 10 nucleotides.

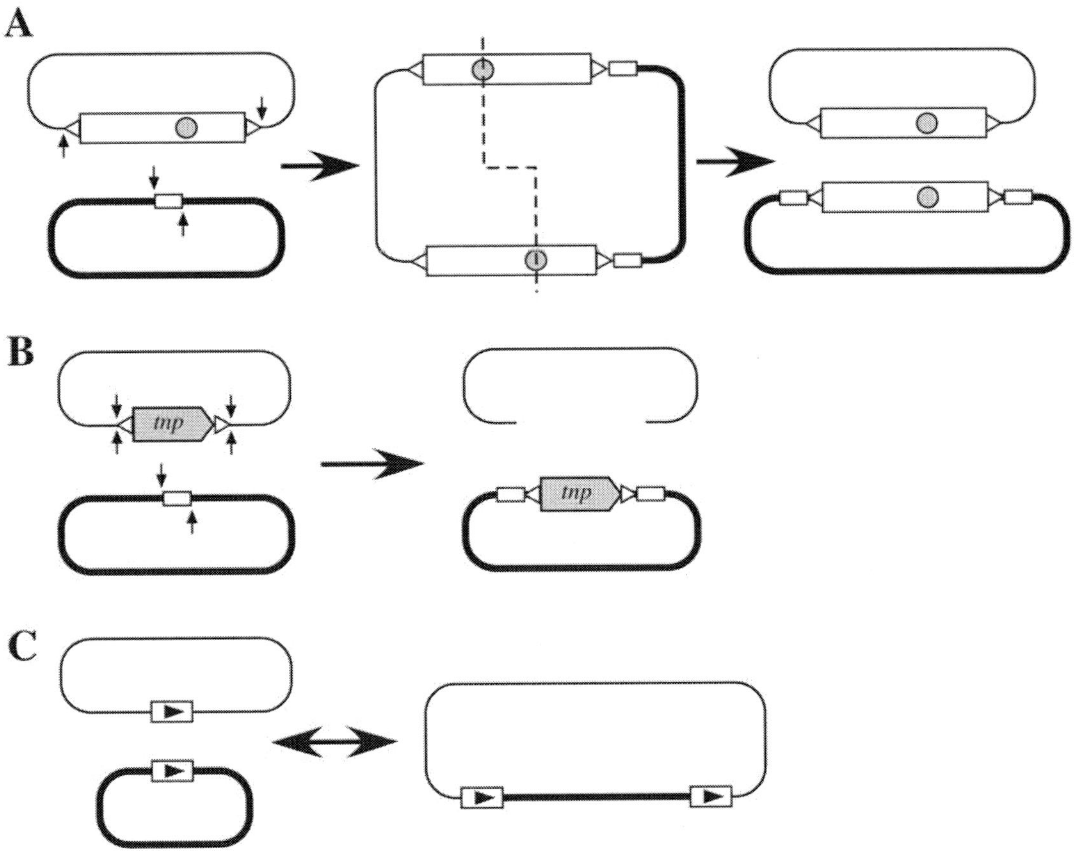

Figure 2. DNA recombination mechanisms mediated by transposons and integrase-based elements. Vertical arrows indicate cleavages by transposase. White triangles show the IRs of transposons, and the white rectangles represent the target sequences. (A) Copy-and-paste transposition used by class II transposons. The 3'-ends of the transposon are cleaved by TnpA and jointed with the staggered ends in the target molecule. The transposon is copied in the reaction, and two copies of the transposon connect the donor and target molecules to form a cointegrate. The cointegrate resolves at the resolution sites (circle) by the function of site-specific recombinase encoded on the transposon (dashed line). As a result, each of the donor and target carries a copy of the transposon. (B) Cut-and-paste transposition used by IS elements. The transposing element is excised from the donor molecule (thin line) because of the double-stranded cleavages at the both ends of the IS, and then integrated into the target molecule (thick line) with a staggered cleavage. The element is thus conserved through the reaction and generates a target duplication upon transposition. (C) Integrase-mediated integration and excision. The integrases mediate site-specific recombination between two attachment sites (rectangle with a black arrowhead), resulting in the integration (right to left) and excision (left to right) of a DNA fragment.

Class II Catabolic Transposons

Toluene-Catabolic Transposons Tn*4651* and Tn*4653*

Transposon Tn*4651* (56 kb) is the first-identified class II transposon that carries the catabolic function [147]. It has the entire genes required for catabolism of toluene/xylenes (*xyl*) into the TCA cycle intermediates (Figure 3) and is loaded on an IncP-9 plasmid, pWW0, identified in *P. putida* mt-2 (Figure 4) [160]. Although transposition of the *xyl*-containing region, *i.e*, Tn*4651*, had already been observed in the late 1970s [45, 48, 89], it was a decade later when a genetic analysis revealed the detailed structure and function of the element [148, 149]. Tn*4651* transposes by the two-step process, cointegration and resolution (Figure 2A). The cointegration reaction is mediated by the TnpA protein that recognizes the 46-bp IRs of the transposon (Figure 1B). There are considerable differences of the IR and *tnpA* sequences between Tn*4651* and other class II transposons such as Tn*3* and Tn*21* (Figure 1B), and the Tn*4651* transposase does not carry out transposition of other elements [149]. The TnpA transposase activity is down-regulated at the post-transcriptional level by the TnpC protein, which is encoded just downstream of the *tnpA* gene (Figure 1A and 4) [44]. The resolution reaction of the Tn*4651*-mediated cointegrate is also specific to Tn*4651*. The resolution system of Tn*4651* is composed of two structural genes, *tnpS* and *tnpT*, and their target site, *res*, which is located between the two genes (Figure 1A and 4) [29]. The *tnpS* and *tnpT* genes are transcribed divergently, and their transcriptional start points have been determined [29]. The TnpS protein is a member of tyrosine recombinases, but TnpT does not show any significant amino-acid sequence similarity to the proteins that are involved in DNA recombination. TnpS can catalyze both site-specific integration and resolution between two *res* sites. TnpT is considered as an enhancer protein upon the resolution reaction, but the exact function of TnpT is still unclear. Tn*4651*-like transposons are also found from other *Pseudomonas* strains: the mercury-resistance transposon Tn*5041* [63] and the carbazole-catabolic transposon Tn*4676* (see below) [73, 125], and both elements have the similar TnpS-*res*-TnpT resolution system (Figure 4B). This might show the concurrent evolution of Tn*4651*-like transposition and resolution systems.

The *xyl* genes in Tn*4651* consist of four transcriptional units, *xyl* upper and lower operons, and two regulatory genes, *xylR* and *xylS* [33, 109]. The upper operon carries *xylUWCMABN* genes (Figure 4A), whose products convert toluene and xylenes to benzoate and its methyl derivatives, respectively. The *xyl* lower operon contains *xylXYZLTEGFJQKIH* (Figure 4A) for further metabolism of benzoate and its methyl derivatives to the TCA cycle intermediates, using the so-called catechol *meta*-cleavage pathway (Figure 3A). Detailed catabolic pathway of the toluene/xylenes is summarized elsewhere [98]. The XylR and XylS regulators control the expression of the upper and lower operons, respectively. The organization of genes within the two *xyl* operons are usually conserved in other toluene/xylenes-degrading plasmids such as pDK1 and pWW53 from two different *P. putida* strains [157, 165], although the relative orientations and copy numbers of the upper and lower *xyl* operons and the incompatibility of the residing plasmids are different from one another (Figure 4A and 5A). These observations suggest that the common ancestral *xyl* operon has been distributed, rearranged, and/or evolved independently in different bacterial populations. All the *xyl* genes in Tn*4651* are sandwiched by two directly repeated copies of IS*1246* [110], forming a 39-kb IS-composite catabolic transposon (Figure 4A). However, the detailed sequence comparison has led to a proposal that the *xyl* genes in Tn*4651* were incorporated not by transposition of the IS*1246*-composite transposon but by intermolecular homologous recombination between two copies of IS*1246*, one on the pWW0 (or Tn*4651*) precursor and the other near the *xyl* genes on another replicon [161]. This idea is consistent with the fact that large IS-composite transposons are hardly transposable. Instead, the two IS*1246* elements on pWW0 mediate intramolecular and reciprocal recombination, generating a 17-kb *xyl*-lacking transposon, Tn*4652* [147].

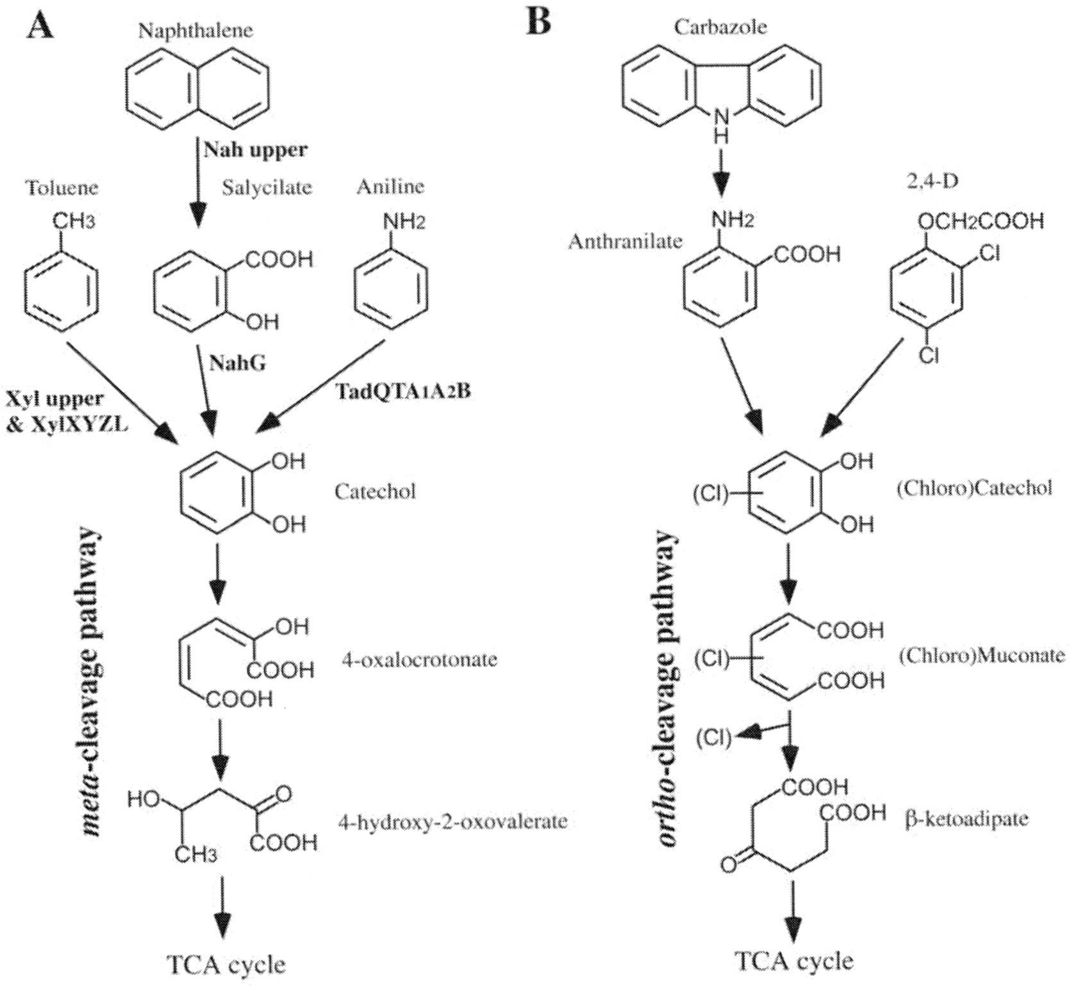

Figure 3. Outlines of catabolic pathways via the *meta*- (A) or *ortho*- (B) cleavage. In each pathway, the target compounds are first converted to catechol or its derivatives, then further metabolized to the TCA-cycle intermediates. See references [40, 98, 111] for more detailed information on these pathways.

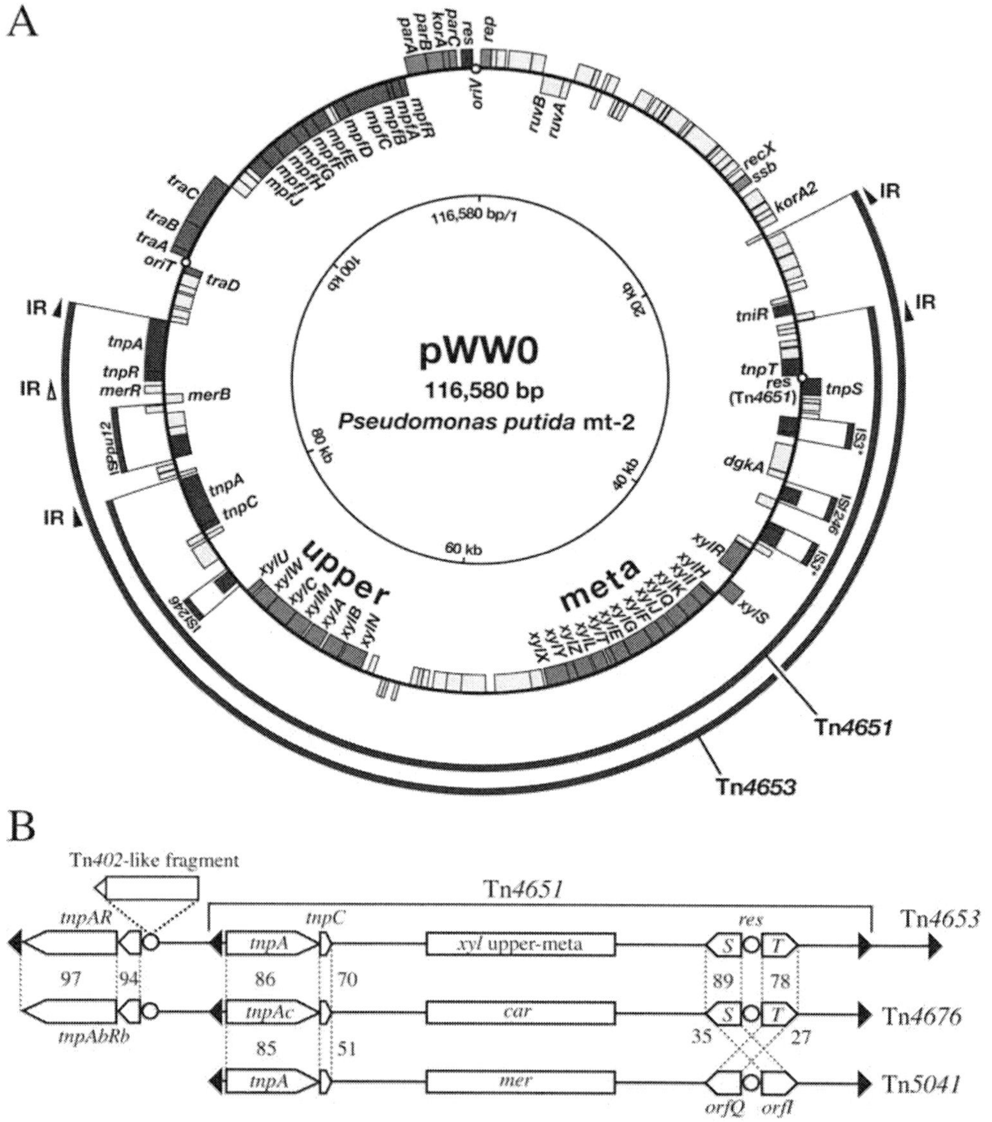

Figure 4. Circular map of pWW0 and schematic linear map of transposons. (A) Map of pWW0. Genes coded in clockwise and counter-clockwise directions are indicated outside and inside, respectively, of the outermost circle. Gene functions are distinguished by the following colors; purple, conjugative transfer; green, replication and partition; blue, transposition and site-specific recombination; red, toluene/xylenes degradation (*meta*, *meta*-cleavage pathway operon, upper, upper pathway operon); and yellow, other functions. See Greated et al. 2002 [33] for the details of the genes on pWW0. Blue arcs indicate transposable regions. 38-bp IRs of Tn*21*-group transposons and 46-bp IRs of Tn*4651*-group transposons are depicted by closed triangle, while the 25-bp IR of Tn*402*- group transposons is shown by an open triangle. The asterisks indicate unnamed insertion sequences belonging to IS*3* family. The sequence information is deposited in GenBank under the accession no. AJ344068. (B) Schematic figures of Tn*4651* and Tn*4653*, and their comparison with related transposons. The triangles are used as shown in the panel A. The pentagons indicate the directions and approximate sizes of orfs. Circles show the *res* sites. The numerals between transposons indicate the amino-acid identities between the corresponding gene products.

Tn*4651* is covered by another class II transposon Tn*4653* with the size of 71 kb (Figure 4) [148, 149]. Tn*4653* carries its own *tnpA* and *tnpR* genes, whose products can complement the *tnpA*-defect of Tn*1722* and the *tnpR*-defects of Tn*1722* and Tn*21*, respectively. Therefore, Tn*4653* is classified into the Tn*1722*-subgroup of the class II transposons. The failure of TnpR to resolve a Tn*4653*-mediated cointegrate is due to an insertion event of a Tn*402*-like transposon into the previously intact *res* site of Tn*4653* (Figure 4B), which is probably caused by the *res*-specific transposition of the Tn*402*-like transposon [81, 133]. Tn*4653* on pWW0 has a 5-bp target duplication at its flanking sequences, indicating the past transposition event of Tn*4653* (or its precursor) into the IncP-9 backbone to generate pWW0. Since pWW0 is a 117-kb plasmid, the 61% portion of the plasmid consists of the transposon-derived sequences.

Toluene-Catabolic Transposons Tn*4656*, Tn*4657*, and Tn*4660*

The IncP-7 plasmid pWW53 is a non-conjugative toluene/xylenes-catabolic plasmid that was identified in *P. putida* MT53 isolated in North Wales (Figure 5) [58, 165]. It was more than twenty years ago that the following facts were shown: (i) the plasmid carries a single copy of the *xyl* upper operon and two homologous but functionally distinguishable *xyl* lower (*meta*-cleavage) operons (*meta* 1 and *meta* 2), (ii) the *xyl* operons are similar to those found in pWW0, and (iii) the *xyl*-containing regions can transpose into other DNA molecules [58, 59, 99].

Moreover, the structures of *xyl* operons as well as their regulation by several regulator proteins have been extensively investigated [3, 28, 58-60, 99]. The approximate structure of a pWW53-borne class II transposon, Tn*4656,* and its transition machinery have also been clarified [152]. However, it was 2007 when the complete sequence of pWW53 was finally determined [165].

Plasmid pWW53 has a 108-kb sequence with very complicated structure. However, its backbone sequence shows high nucleotide similarity to other completely sequenced IncP-7 plasmids, the carbazole-catabolic plasmid pCAR1 and the naphthalene-catabolic plasmid pND6-1 [70, 73]. Since pCAR1 carries a whole set of its transfer genes (Figure 7, see below) [73], the transfer machinery of pWW53 is thought to have lost in its evolutionary steps. A noteworthy structural feature of this plasmid is that it carries two distinct transposase-resolvase (serine recombinase) gene clusters (*tnpRA* modules), five IRs, and three *res* sites, which are all typical of class II transposons. These genetic materials lead to the formation of four possible transposable regions, Tn*4657* to Tn*4660* (Figure 5). The largest 86-kb region (Tn*4657*) occupies 80% of the pWW53 sequence, and covers the three other transposons. Intriguingly, the replication-associated genes of pWW53 are also carried on Tn*4657*, suggesting that any DNA molecules with the insert of this transposon can behave as IncP-7 replicons. Of the four possible transposons, Tn*4657* and Tn*4660* (62 kb) cover all of the 36 *xyl* genes, and Tn*4657*, Tn*4658* (5.7 kb), and Tn*4659* (3.7 kb) are still transposable. Although Tn*4656* is the 37-kb first-identified toluene-catabolic transposon from pWW53, this transposon is not present on pWW53 as a contiguous segment. The components of Tn*4656* are carried as two separated and inversely oriented segments (3.7 and 33 kb) (Figure 5). The 3.7-kb region is an IR2-*tnpAR-res* part of Tn*4658*, and the 33-kb one shares the IR1 end with Tn*4657* (Figure 5). The Tn*4658* resolvase is a serine-recombinase, and Tn*4656* is formed by the Tn*4658* resolvase-mediated site-specific inversion of a large pWW53 segment between the two inversely oriented *res* sites, *res1* and *res2* on pWW53 (Figure 5B). The Tn*4658* transposase is able to mediate the transposition of Tn*4658*, Tn*4657*, and Tn*4656*, while the Tn*4659* transposase catalyzes only the transposition of Tn*4659*. Tn*4660* on pWW53 has at its flanking regions the 4-bp duplicated sequences, which strongly suggest its past transposition event into the ancestral form of pWW53. However, transposition of Tn*4660* was not detected in the presence of either transposases, probably because of the mutated IR1 sequence by the insertion event of IS*1492*. Replacement of the IR1 with the putative ancestor of IR1 (IRb) led to the transposition of Tn*4660* with concomitant generation of 4- or 5-bp target duplication. At present, the plausible explanation for the establishment of pWW53 is the transposition of Tn*4660* into an IncP-7 backbone sequence. Tn*4658* must have been acquired some time during the evolution of the plasmid. These findings clearly indicate that multiple copies of transposition-related genes and sites on one plasmid and their recombination activities contribute greatly to the diversification of plasmid structures.

Naphthalene-Catabolic Transposon Tn*4655*

Plasmid NAH7 from *P. putida* G7 is a self-transferable plasmid with all genes required for the naphthalene degradation (*nah*) (Figure 6) [24]. NAH7 was assigned to the IncP-9 group of plasmids before its nucleotide sequence became available because of (i) its failure to coexist with other IncP-9 plasmids [159] and (ii) its close relationship with pWW0 with respect to the backbone sequences and degradation genes (see above) [4, 33, 39, 68]. The degradation cascade of naphthalene and its responsible genes and proteins have been studied extensively as a model of the bacterial PAH degradation [39, 79, 119, 166, 167]. It has been demonstrated that the *nah* genes are carried on a defective class II transposon, Tn*4655* (Figure 6) [150]. We determined in 2006 the complete nucleotide sequence of NAH7 to reveal the structures and functions of the plasmid and transposon [132].

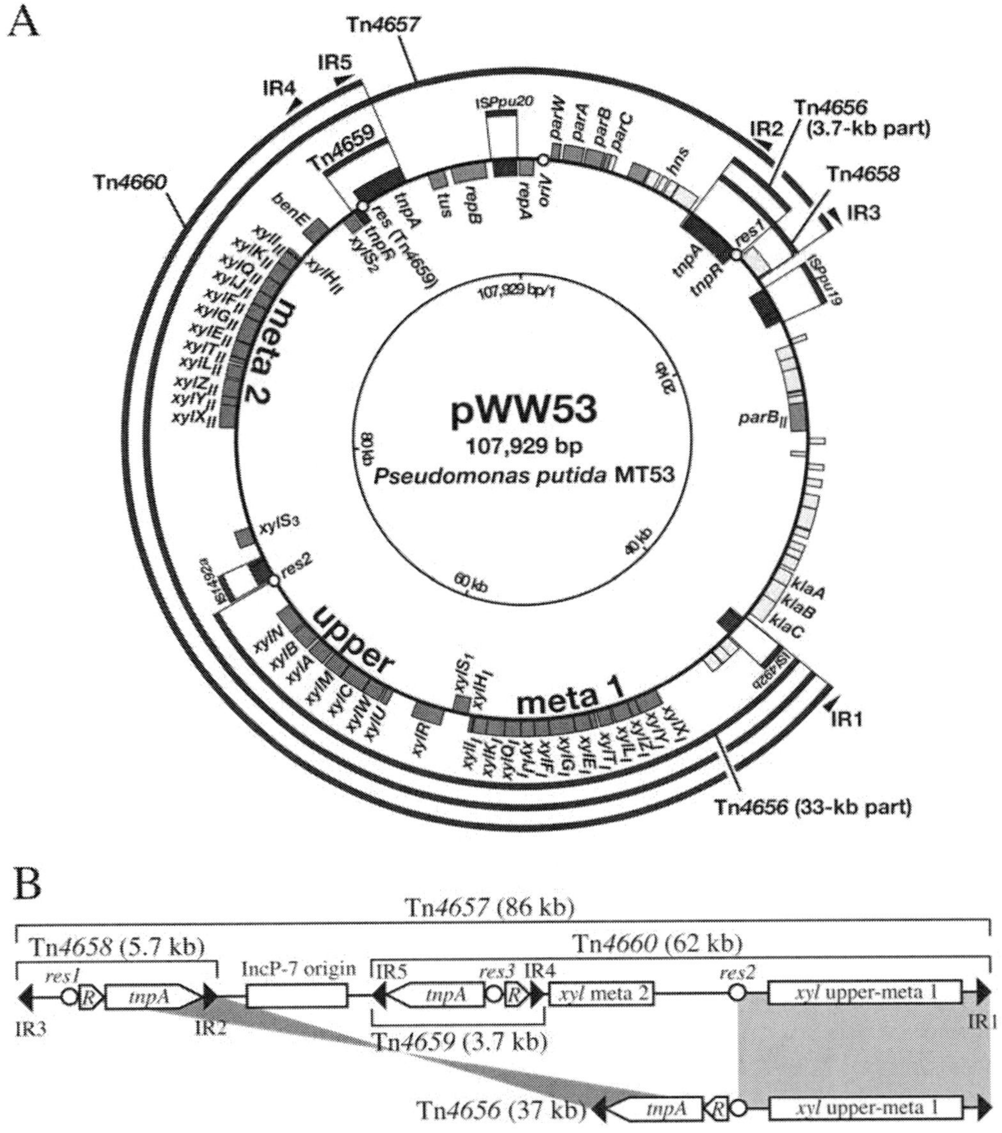

Figure 5. Circular map of pWW53 and schematic linear maps of transposons. See the legend of Figure 4 for the explanation of genes, their functions, IRs, etc. (A) Circular map of pWW53 showing five class II transposons, Tn*4656* to Tn*4660*. The *res1* and *res2* sequences, which are highly homologous and inversely oriented, are resolution sites catalyzed by Tn*4658* TnpR to generate Tn*4656* (see text for details). The sequence information is deposited in GenBank under the accession no. AB238971. The details of the genes on pWW53 are described in Yano et al. 2007 [165]. (B) Schematic figures of five transposons. Tn*4657* covers Tn*4658*, Tn*4659*, and Tn*4660*. The site-specific recombination between *res1* and *res2* catalyzed by the Tn*4658* TnpR generates Tn*4656*. Identical regions are shown by light (same strand) or dark (different strand) shadings.

The *nah* genes in NAH7 are organized in two major segments: the *nah* upper- and lower (*meta*)-pathway operons (Figure 6). The gene products encoded by the upper-pathway operon (*nahAa* to *nahD*) convert naphthalene to salicylate [6, 167]. The enzymes encoded by the *nah* lower operon (*nahG* to *nahJ*) convert salicylate to pyruvate and acetaldehyde via the catechol *meta*-cleavage pathway [7, 167]. The details of NAH7-specified naphthalene degradation have been reviewed elsewhere [6, 7, 31, 95, 98, 167]. The product of *nahY* gene in the lower *nah* operon is involved in chemotaxis to naphthalene [34], while the function of *nahX* gene is unknown. A LysR-type transcriptional regulator NahR controls the expression of the two operons [166]. There are many naphthalene- and/or phenanthrene-degrading *Pseudomonas* strains that carry, on their chromosomes or plasmids, the degradation operons very similar to those on NAH7 [6, 7, 20, 64, 70, 137], indicating that those operons have evolved from a common ancestor. It should also be noted that most parts of the *meta*-cleavage pathway operons for aromatic compounds (toluene, naphthalene, aniline etc.) usually have similar gene organizations. On the other hand, the upper-pathway operons and the first small parts of the *meta*-cleavage pathway operons for the conversion of target substrates to catechols (Figure 3A) are totally distinguishable. These observations strongly suggest that the degradability of aromatic compounds through the *meta*-cleavage pathway have been acquired by combining the unique upper-pathway operons and the first small parts of *meta*-cleavage pathway operons to the common parts of *meta*-cleavage pathway operons in the same cells (Figure 3A).

Tn*4655* carries all *nah* genes between 38-bp IRs (IR1 and IR6), which are similar to those of the Tn*1722*-subgroup transposons (Figure 6B). It also has four additional copies of IR-like sequences (IR2 to IR5) that show high similarity with the IRs of other class II transposons. IR5 and IR6 are similar to those of Tn*4653* on pWW0, and both ends of the intervening segment (9.6 kb) also show high (>90%) similarity to the partial sequence of Tn*4653* (Figure 6B). This indicates that the IR6-proximal end of Tn*4655* shares a common evolutionary origin with Tn*4653*. An internal portion of the 9.6-kb internal fragment carries the *attI-orf24-tnpI* cluster that is more than 85% identical to the corresponding sequence on the IncP-7 carbazole-catabolic plasmid pCAR1 (Figure 6B) (see the Tn*4676* section) [73]. The function of the *attI-orf24-tnpI* cluster is described below. The region bracketed by IR3 and IR4 shows very high similarity to a mercury-resistance class II transposon, Tn*5046* [82] (Figure 6B), but lacks its transposition-related genes. These observations indicate that Tn*4655* must have been created after several rearrangements mediated by multiple class II transposons. Since Tn*4655* on NAH7 has a 5-bp target duplication at its flanking sequences, this unit had most likely transposed to the IncP-9 backbone, leading to generation of NAH7. This idea is supported by the facts that (i) The total 36.4-kb regions of NAH7 except for Tn*4655* are conserved in other IncP-9 plasmids, pDTG1 [20] and pWW0 [33], (ii) the backbone of NAH7 showed only 70% identity with pWW0, and (iii) all these three plasmids have insertion of catabolic gene clusters at different positions in their backbones [132].

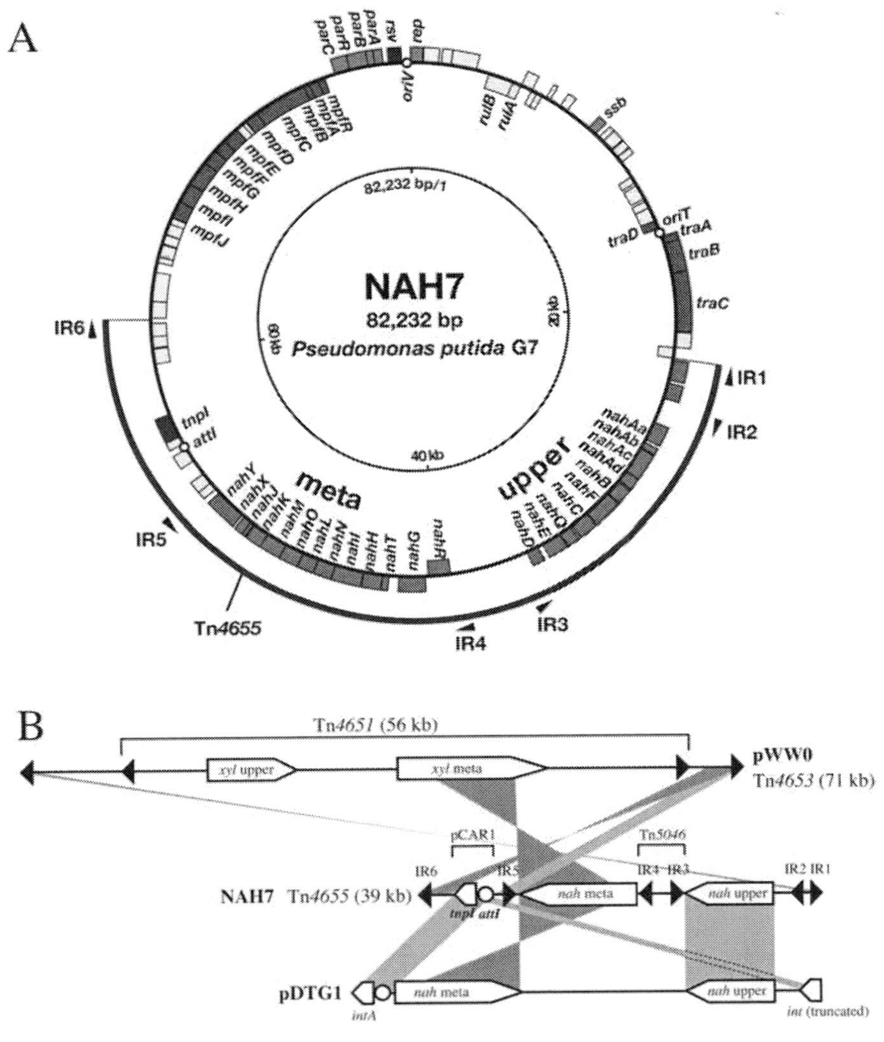

Figure 6. Circular map of NAH7 and schematic linear map of Tn*4655*. See the legend of Figure 4 for the explanation of genes, their functions, IRs, etc. except that the genes for the degradation of naphthalene are depicted in red. The details of the genes on NAH7 are described by Sota et al. [132]. (A) Circular map of NAH7. Transposition of Tn*4655* into the IncP-9 backbone would have generated NAH7. The sequence information is deposited in GenBank under the accession no. AB237655. (B) Schematic figure of Tn*4655* and its comparison with the related DNA fragments. Homologous regions are shown by light (same strand) or dark (different strand) shadings. See text for the detailed information.

Tn*4655* carries its specific resolution system but lacks its own transposase gene. The transposition-defect of Tn*4655* is complemented by supply of the transposase of a Tn*1722*-subgroup transposon [150], and therefore, Tn*4655* can be categorized into the Tn*1722* subgroup. In contrast, the Tn*4655* resolution system is completely different from those of the Tn*1722*-subgroup transposons, and the difference has been confirmed by the sequence analysis of NAH7. Whereas the known Tn*1722*-subgroup transposons adopt the serine recombinase-based resolution system, the Tn*4655* resolution system is mediated by a tyrosine-recombinase (integrase), TnpI, and its recognition site, *attI* [17, 132]. The resolution/integration reaction occurs within the 119-bp *attI* sequence [132]. Between *attI* and *tnpI* is another gene, *orf24*, whose function is unknown. A noteworthy finding is that the homologues of the Tn*4655 attI-orf24-tnpI* module are also present on other catabolic plasmids, the naphthalene-catabolic plasmids pDTG1 (IncP-9) [20] and pND6-1 (IncP-7) [70], and the carbazole-catabolic plasmid pCAR1 (IncP-7) [73], but not on any other molecules in the databases. Therefore, it is possible to speculate that these TnpI integrases have involved in the incorporation of catabolic gene clusters into plasmids. This is consistent with the observations that (i) the two *nah* operons and their intervening sequences on pDTG1 (IncP-9) and pND6-1 (IncP-7) are almost identical to each other, and (ii) the segment on pDTG1 containing all the *nah* genes is bracketed by two copies of the directly repeated *tnpI*-like gene (*intA*), one copy with truncation (Figure 6B). However, it remains unknown how catabolic gene clusters and integrase genes have been introduced into class II transposons to generate large catabolic class II transposons.

Carbazole-Catabolic Transposon Tn*4676*

Carbazole (CAR) is an environmental pollutant rich in coal tar and shale oil, and has a heterocyclic aromatic structure containing a dibenzopyrrole structure (Figure 3B). CAR is a structural homologue of dibenzo-*p*-dioxin (DD) and dibenzofulan (DBF), and is known to possess strong mutagenicity and toxicity [1]. *P. resinovorans* CA10 isolated from activated sludge can utilize CAR as a sole source of carbon, nitrogen, and energy [100]. The genes for the CAR degradation, the *carAaAaBaBbCAcAdDFE* and *antABC* clusters, have been identified on the CA10-residing plasmid pCAR1 (Figure 7) [73, 93]. Plasmid pCAR1 is classified into the IncP-7 group on the basis of its structure and function of replication machinery, and can replicate and transfer within *Pseudomonas* strains [73, 124, 126]. The CarAaAcAd, CarBaBb, and CarC proteins from the former cluster are required for the conversion of CAR to anthranilate and 2-hydroxypenta-2,4-dienoate, the latter of which is further metabolized to pyruvate and acetyl CoA by the CarD, CarE, and CarF proteins [93, 117, 118]. The *antABC* products convert anthranilate to catechol (Figure 3B) [93, 117, 118]. The enzymes to metabolize catechol to TCA cycle intermediates through the β-ketoadipitate (=catechol *ortho*-cleavage) pathway (Figure 3B) are encoded by the CA10 chromosome. The CarAaAcAd complex, namely carbazole 1,9a-dioxygenase (CARDO), exhibits a broad range for its substrates such as CAR, DD, DBF, naphthalene, anthracene, fluorine, fluoranthene, and biphenyl, and can catalyze diverse oxygenation: angular dioxygenation, *cis* dihydroxylation, and monooxygenation [92]. Since the initial dioxygenation of aromatic compounds is the key reaction to determine the degradability and pathway of the compounds [55, 95], the broad substrate range of CARDO is helpful for the initial oxygenation of various aromatic environmental pollutants.

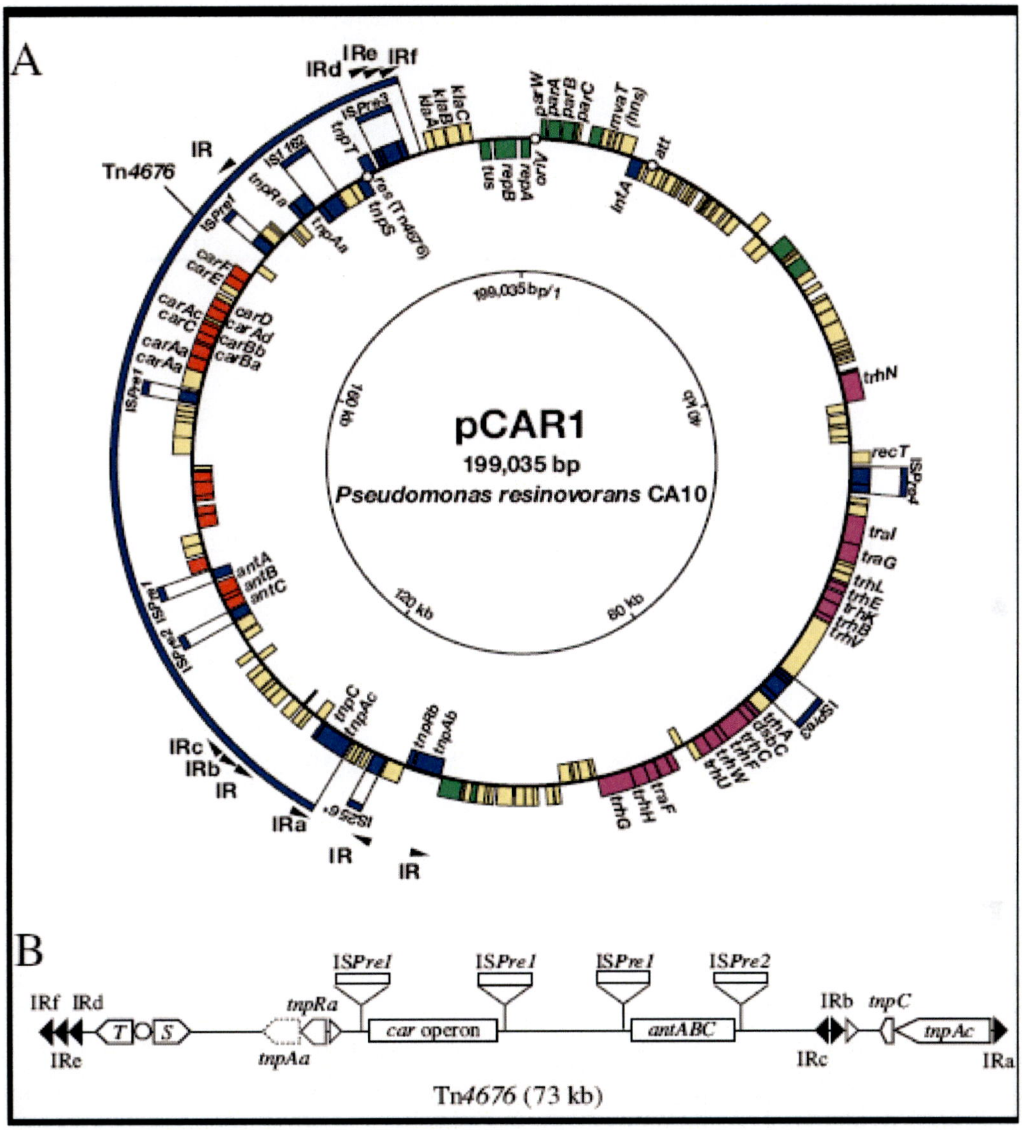

Figure 7. Circular map of pCAR1 and schematic linear map of Tn*4676*. The figure is drawn in the manner the same as that in Figure 4 except that the genes depicted in red are anthranilate (*ant*) and carbazole (*car*) degradation. (A) Circular map of pCAR1. Asterisk, unnamed insertion sequence belonging to IS*256* family. The sequence information is obtained from GenBank accession no. AB088420. See Maeda et al. 2004 [73] for the details of the genes on pCAR1. (B) Schematic figure of Tn*4676*. The black and white triangles indicate the 46- and 38-bp IRs of Tn*4651*- and Tn*21*-group transposons, respectively. The *tnpAa* gene is truncated. Comparison of Tn*4676* with the cognate transposon Tn*4651* is shown in Figure 4B.

The *car* and *ant* clusters are loaded on a 73-kb class II transposon, Tn*4676* (Figure 7) [73, 125]. Tn*4676* has the 46-bp terminal IRs (IRa and IRf), a transposase gene (*tnpAc*), and a resolution system (*tnpS-res-tnpT*) that show high (>70%) similarity to those of the toluene-catabolic transposon Tn*4651*, indicating that it belongs to the Tn*4651*-group of class II transposons. Similarly to Tn*4651*, Tn*4676* has the *tnpC* gene just downstream of the *tnpAc* gene (Figure 4B and 7). The transposon is bracketed by 5-bp doublet sequences, indicating that past transposition event of Tn*4676* resulted in the formation of pCAR1. Tn*4676* also carries four additional copies of the 46-bp IR (IRb to IRe), and two copies of the 38-bp IR and a *tnpRA* module (*tnpRaAa* with truncation of *tnpAa*) that are similar to those of Tn*4653* (Figure 7), indicating repeated molecular rearrangements associated with class II transposons. Four nearly identical copies of IS elements (three IS*Pre1* and one IS*Pre2*) are also located on Tn*4676* [93]. The *ant* genes are situated between IS*Pre1* and IS*Pre2*, and the *car* genes are sandwiched by IS*Pre1* (Figure 7B). It is uncertain whether these two gene clusters had been incorporated as IS-composite transposons into Tn*4676*. Outside of Tn*4676* is another *tnpRA* module (*tnpRbAb*) and two 38-bp IRs that are also similar to that of Tn*4653* (Figure 7A). This finding implies that Tn*4676* might be located in another Tn*4653*-like transposon. However, no 38-bp IR sequences are found at the other outside region of Tn*4676* on pCAR1. Although the relative orientations and nucleotide sequences of the *tnpRbAb* and *tnpAcC* modules of Tn*4676* on pCAR1 are similar to those of the *tnpRA* and *tnpAC* modules in Tn*4653* on pWW0 (Figure 4B), the intervening sequences between the two modules are quite different from one another. Therefore, it is likely that (i) Tn*4676* was located or inserted in a Tn*1722*-like transposon in the past and (ii) the transposons on pCAR1 and pWW0 have evolved independently.

Haloacetate-Catabolic Transposon Tn*Had2*

A soil bacterium *Delftia acidovorans* strain B was isolated from industrial wastewater in Japan, and can utilize monohaloacetates as sole sources of carbon and energy [57]. Two monohaloacetate dehalogenases, H-1 and H-2, were identified in this strain. H-1 acts predominantly on fluoroacetate, and H-2 on chloro-, bromo-, and iodoacetate [56]. The strain also shows resistance to mercury [57]. The two dehalogenase genes, *dehH1* for H-1 and *dehH2* for H-2, as well as the mercury-resistance genes (*mer*) are carried on the 67-kb IncP-1β plasmid pUO1 (Figure 8). Sequence analysis of pUO1 revealed that the *dehH2* gene is flanked by two intact and identical copies of IS*1071*, therefore leading to the structure of an IS*1071*-composite transposon named as Tn*Had1* (Figure 8) [129]. The *dehH1* gene is also bracketed by two copies of IS*1071*, one shared with Tn*Had1*, but the other is truncated (Figure 8B) [129]. This indicated that the *dehH1* gene was also carried on an IS*1071*-composite transposon in the past. This truncation of the Tn*Had1*-distal IS*1071* copy was probably generated by the past insertion of a Tn*21*-like transposon into this IS*1071* [129]. Detailed sequence analysis surrounding the three copies of IS*1071* and two *deh* genes have revealed the presence of three Tn*21*-like IR sequences (IR1 to IR3), two of which are situated at the junctions between the IncP-1β plasmid backbone and the 15.6-kb *deh*-containing segment (Figure 8). The segment designated Tn*Had2* is flanked with 5-bp duplicated sequences, indicating that Tn*Had2* had transposed into the ancestor of pUO1 to establish the present structure. However, no Tn*21*-like transposition-related genes are located in Tn*Had2* except for the *res* site. Complementation analysis by supply of the *tnpA* and *tnpR* genes of Tn*21* or Tn*1722* in *trans* have shown that Tn*Had2* is transposable by the Tn*21* transposase but not by the Tn*1722* one, and that the resolution between the Tn*Had2 res* sites is mediated by both the Tn*21* and Tn*1722* resolvases.

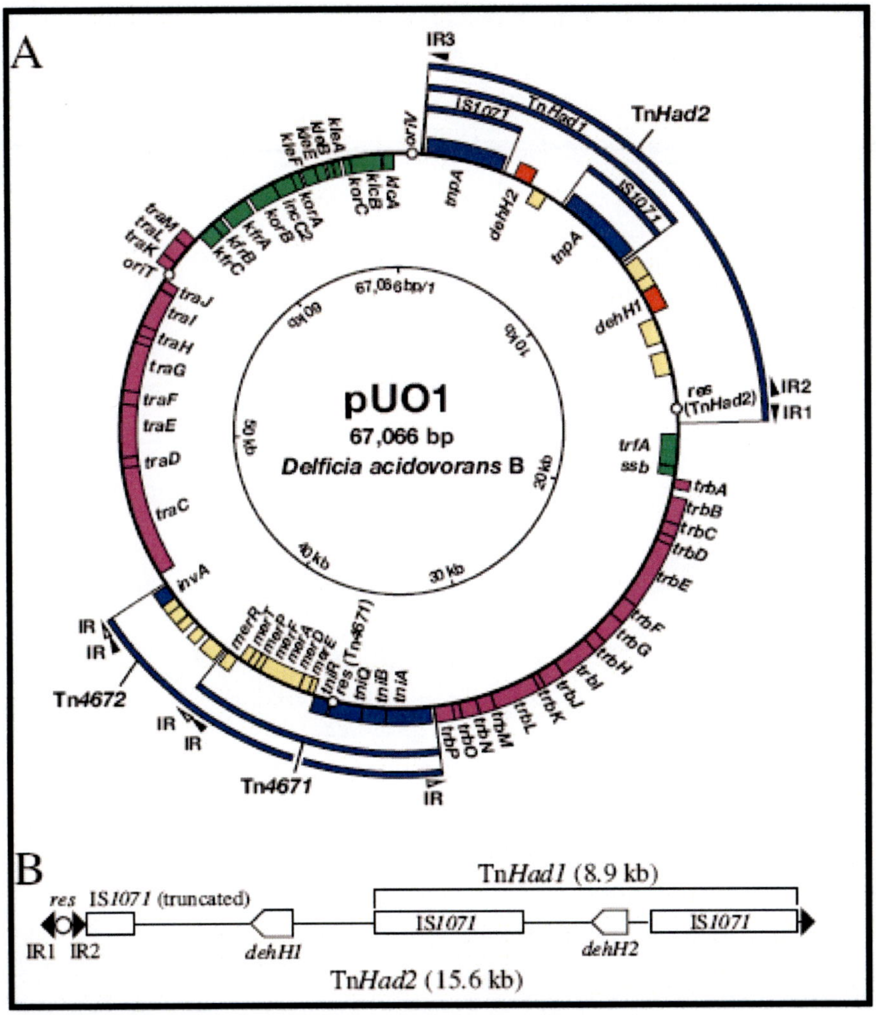

Figure 8. Circular map of pUO1 and schematic linear maps of transposons. The figure is drawn in the manner the same as that in Figure 4 except that the genes depicted in red are for haloacetate dehalogenation. (A) Circular map of pUO1. The regions where Tn*Had2* and Tn*4672* are situated are the two typical ones for the insertion of foreign DNA fragments in IncP-1 plasmids [121, 133] (Figure 9). The sequence information is deposited in GenBank under the accession no. AB063332. (B) Schematic figures of Tn*Had1* and Tn*Had2*. TnHad1 is an IS1071-composite transposon with the *dehH2* gene. The IS*1071* sequence downstream of the *dehH1* gene is truncated because of the insertion of a Tn*21*-like IR (IR2).

From these results, Tn*Had2* is classified as a class catabolic II transposon belonging to the Tn*21*-subgroup in the Tn*21* family.

A noteworthy outcome from the Tn*Had2* study is the characterization of the outermost GGGG sequences conserved in the IRs of all the known class II transposons (Figure 1B). The G-to-C mutations in any of these four nucleotides, even in one IR, result in the decreases in transposition frequencies of the mutated Tn*Had2* derivatives in the presence of the Tn*21* TnpA protein [129, 130] (Table 2).

Table 2. Transposition of the Tn*Had2* derivatives with the mutant end(s)*

Terminal nucleotides§	Relative transposition frequency†
5'-GGGG + 5'-GGGG	1.0
5'-GGGG + 5'-AGGG	1.6×10^{-1}
5'-GGGG + 5'-CGGG	1.6×10^{-1}
5'-GGGG + 5'-TGGG	4.6×10^{-1}
5'-CGGG + 5'-CGGG	1.3×10^{-1}
5'-GAGG + 5'-GAGG	1.2×10^{-3}
5'-GTGG + 5'-GTGG	$< 7.6 \times 10^{-7}$
5'-GCGG + 5'-GCGG	$< 5.3 \times 10^{-7}$
5'-GGGG + 5'-GCGG	1.8×10^{-2}
5'-GGCG + 5'-GGCG	2.0×10^{-3}
5'-GGGC + 5'-GGGC	6.5×10^{-3}
5'-CGCC + 5'-CGCC	3.6×10^{-4}

* Transposition of the Tn*Had2* derivatives was investigated in the presence of the transposition genes of Tn*21*.

§ Only the four outermost nucleotides of the IRs are shown for simplicity.

† The transposition frequency of Tn*Had2* flanked by the two 5'-GGGG ends (top row) is taken as 1.0 to calculate the relative transposition frequency. The average transposition frequency at least from three independent experiments is shown.

While the mutation at the first position (outermost end) of both IRs lead to slight (10-fold) decrease in the transposition frequency, the mutation at the second, third, or fourth positions cause the 150- to 1000-fold reduction of transposition frequencies. The critical nucleotide is the second one, because its mutation to pyrimidine (C or T) nucleotide abolishes the transposition of the Tn*Had2* derivatives [130] (Table 2). The G-to-A mutation at the second position allows the transposition at a 1000-fold lower frequency. In addition, a Tn*Had2* derivative with the CGCC sequences in both IRs is still transposable, albeit at the frequency 2,800-fold lower than of Tn*Had2* (Table 2). These results suggest that the Tn*21* TnpA protein (i) binds tightly to the second to fourth G nucleotides and (ii) recognizes a purine nucleotide at the second position of the 38-bp IRs. The latter point could be supported by a recently published sequence of the IncP-1 plasmid pA81 from *Achromobacter xylosoxidans* (accession no. AJ515144) in which a class II transposon, Tn*AxII* [50], has the outermost GAGG sequences in its IRs. More detailed biochemical studies are necessary to clarify detailed interaction between TnpA and IRs.

Complete nucleotide sequencing of pUO1 revealed that the *mer* genes are located on two Tn*402*-like transposons, Tn*4671* and Tn*4672*, which share one end (Figure 8) [130]. Tn*402*-like transposons are relatively newly established group of transposons that require 25-bp IRs and four genes to mediate transposition [62, 134]. Tn*4671* is nearly identical to the mercury-resistance transposon Tn*5053* [61]. Tn*402*-like transposons are known to specifically transpose into the *res* sites for the resolvases in canonical class II transposons [81], and the Tn*4671* transposition into the *res* site of Tn*Had2* has been indeed detected [130]. Our recent study has demonstrated that the target specificity of Tn*402*-like transposons as well as the region-specific transposition of Tn*21*-like transposons are the cause of the structural similarity of the IncP-1 plasmids (Figure 9) [133]. Since IncP-1 plasmids carry the resolvase genes and *res* sites between the *tra* and *trb* operons, Tn*402*-like transposons in the known IncP-1 plasmids are found in this inter-operonic region, the *tra-trb* region (Figure 9). In contrast, Tn*21*-like transposons prefer to transpose between the replication-initiation protein gene, *trfA*, and its *cis*-acting site, *oriV* [133] (Figure 9), and many Tn*21*-like transposons and their remnants are indeed identified in this "*oriV-trfA* region". Therefore, most IncP-1 plasmids have two typical insertion sites for two species of mobile elements and show similar basic structure regardless of the differences in subgroups (α to ε) and in phenotypic traits [121]. These facts suggest that IncP-1 plasmids, including pUO1, have evolved by acquisition of transposons at one or both of the specific regions in their transposon-free ancestors such as pBP136 and pA1 [38, 54]. All of IncP-1 catabolic plasmids so far reported have IS*1071* sequences in the vicinity of the catabolic gene clusters, and IS*1071* has not been identified on any other groups of plasmids from G^- bacteria except for IncP-1 plasmids (Table 3). This particular partnership between IS*1071* and IncP-1 plasmids is a key to better understand the generation of IncP-1 catabolic plasmids. The intimate relationship between IS*1071*, catabolic genes, and IncP-1 plasmids are further described below.

Table 3. Distribution of IS*1071* in bacterial genomes

Bacteria	Bacterial division*	Location of IS*1071*§	Inc group of plasmid †	Phenotype¶	Reference
Gram-negative					
Achromobacter xylosoxidans A8	β	pA81	P-1β	2-Chlorobenzoate	[49]
A. xylosoxidans EST 4002	β	pEST4011	P-1δ	2,4-D	[156]
Acidovorax sp. JS42	β	Chr, pAOBO02	P-1β	2-Nitrotoluene	[69]
Bordetella bronchiseptica	β	R906	P-1β	Streptomycin	[158]
Burkholderia cepacia AMMD	β	Plasmid 1	P-1β	N/D	[141]
B. cepacia 2a	β	pIJB1	P-1β	2,4-D	[105]
B. xenovorans LB400	β	Chr		N/D	[11]
Comamonas sp. CNB-1	β	pCNB-1	P-1β	4-Chloronitrobenzene	[72]
C. teststeroni T-2	β	pTSA	P-1β	p-Toluenesulfonate	[145]
C. testosteroni BR60	β	pBRC60	P-1β	3-Chlorobenzoate	[87]
Cupriavidus eutrohpa JMP134	β	pJP4	P-1β	2,4-D	[146]
C. metallidurans CH34	β	Chr, pMOL28	N/A	N/D	[80, 84]
D. acidovorans	β	Chr		Diethylphosphate	[138]
Delftia acidovorans B	β	pUO1	P-1β	Haloacetate	[129, 130]
D. acidovorans P4a	β	Chr		2,4-D	[43]

Table 3. (Continued)

Bacteria	Bacterial division*	Location of IS*1071*§	Inc group of plasmid†	Phenotype¶	Reference
D. tsuruhatensis AD9	β	Chr		Aniline	[71]
Pseudomonas sp. ADP	γ	pADP-1	P-1β	Atrazine	[77]
P. pavonaceae 170	γ	Chr		1,3-dichloropropene	[104]
P. pseudoalcaligenes POB310	γ	pPOB	N/A	Carboxydiphenyl ether	[19]
P. putida UCC22	γ	pTDN1	N/A	Aniline	[26]
Ralstonia sp. U2	β	pWWU2	N/A	Naphthalene	[169]
Gram-positive					
Arthrobacter aurescens TC1	Actinobacteria	Chr, pAA1, pTC1	N/A	atrazine	[85, 116]

* α, β, and γ, α-, β-, and γ-Proteobacteria, respectively.
§ Chr, chromosome.
† N/A, not assigned.
2,4-D, 2,4-dichlorophenoxyacetic acid; N/D, not determined.

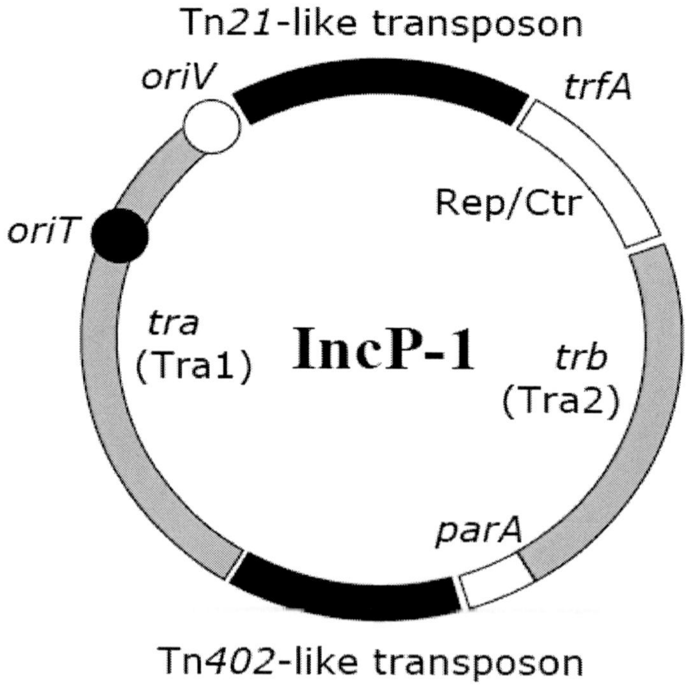

Figure 9. Schematic structure of IncP-1 plasmid. Abbreviations: *trfA*, replication-initiation protein gene; Rep/Ctr, region required for plasmid replication, maintenance, and control; *tra* and *trb*, operons to mediate plasmid transfer; *parA*, resolvase gene; *oriT*, origin of plasmid transfer; and *oriV*, origin of vegetative plasmid replication. Tn*21*-like and Tn*402*-like transposons are mainly situated between *oriV* and *trfA* and between *tra* and *trb* operons, respectively.

Catabolic Genes Associated with Class II Insertion Sequence IS*1071*

IS*1071* is the only IS element belonging to the class II transposons. This element was first found as the IS components of the chlorobenzoate-catabolic composite transposon, Tn*5271* [87]. Its TnpA protein shows only 20% identity with TnpA proteins of other class II transposons, and the 110-bp IRs are much longer than those of others. Various deletion mutations in the 110-bp IRs result in the loss of transposition of the IS*1071* derivatives, strongly suggesting that most, if not all, of the 110-bp sequence is required in *cis* for transposition [131]. This transposon is also the only class II transposon that intrinsically does not encode any resolution function. Thus, IS*1071* is significantly different from other class II transposons in the structure, function, and phylogeny. Since class II transposons are considered to have evolved by acquisition of various resolution functions, IS*1071* might have been an ancestral form of class II transposons [36, 131, 163]. A noteworthy feature of the transposon is that most IS*1071* elements are (i) located in the vicinity of genes for degradation of xenobiotic compounds (Figure 10), (ii) situated mainly on IncP-1 plasmids, and (iii) found predominantly in β-proteobacterial strains (Table 3). The predominant distribution in β-proteobacteria is consistent with our finding that it is only transposable in this phylum [131]. These observations strongly suggest an important role of IS*1071* in the dissemination of xenobiotic degradation genes in conjunction with the broad-host-range HGT event of IncP-1 plasmids.

Although the transposons bracketed by two copies of IS*1071*, including Tn*5271*, are structurally categorized into class I-composite transposons (Figure 1), their transposition mechanism is basically similar to that of class II transposons except the lack of resolution event. Therefore, the IS*1071*-composite transposons can be regarded as class II-composite transposons. The resolution events can sometimes be compensated for by the homologous recombination between two copies of IS*1071*. Generation of IS*1071*-composite transposons may be dependent on the transposition mechanism specific to IS*1071*. Wyndham et al. [163] have proposed a model for the formation of IS*1071*-composite transposons (Figure 11). Firstly, a donor molecule with an IS*1071* element forms, upon its transposition, a stable cointegrate with the target molecule (Figure 11B). A copy of IS*1071* is duplicated at this step, and each IS*1071* locates at the junctions of the donor and target molecules (Figure 11). A 5-bp duplication of the target sequence is also generated at the junctions. The duplication of IS*1071* and the 5-bp target sequence has been experimentally confirmed by our study [131]. Subsequent intramolecular transposition of one IS*1071* copy to a novel target site on the other side of a catabolic gene cluster results in the deletion of the initial donor molecule from the cointegrate (Figure 11C). Consequently, the donor molecule acquires an IS*1071*-composite transposon carrying the gene cluster, and the target molecule gets a deletion of the cluster, leaving a copy of IS*1071* (Figure 11C). This model explains well why large catabolic gene clusters can be incorporated into plasmids as IS*1071*-composite transposons. Since the 5-bp duplications generated by the two transposition events are separately located on the donor and target molecules, no copy of IS*1071* has the same 5-bp sequences outside of the element (Figure 11C). This is consistent with the fact that most of IS*1071* found so far do not have such sequences. Although the detailed transposition mechanism of IS*1071* remains to be elucidated, it has obviously been involved in dissemination of catabolic genes for xenobiotics. Below are examples of the genes possibly propagated and disseminated by IS*1071*.

3-Chlorobenzoate Degradation

A β-proteobacterial strain, *Comamonas testosteroni* BR60 (formerly *Alcaligenes* sp. BR60), was isolated from a chlorobenzoate-contaminated landfill, and can grow on 3-chlorobenzoate [162]. The genes required for the 3-chlorobenzoate catabolism (*cbaABC*) are carried on the 88-kb plasmid pBRC60, and bracketed by two copies of IS*1071* to form a 17-kb composite transposon Tn*5271* (Figure 10). Plasmid pBRC60 is considered as an IncP-1β plasmid on the basis of its structural similarity to the archetypal IncP-1β plasmid R751 [140] and hybridization data using several IncP-1β–specific genes as probes [9].

Tn*5271* and the mercury-resistance genes (*mer*) in pBRC60 are located at the positions similar to Tn*Had2* and Tn*4672*, respectively, on pUO1. The *cbaABC* genes in Tn*5271* encode 3-chlorobenzoate-3,4/4,5-dioxygenase (*cbaA*), 3-chlorobenzoate-3,4/4,5-dioxygenase reductase (*cbaB*), and 1-carboxy-3-chloro-3,4-dihydroxycyclohexa-1,5-diene dehydrogenase (*cbaC*), which sequentially convert 3-chlorobenzoate to protocatechuate [88]. The protocatechuate thus produced is metabolized via the protocatechuate *meta*-cleavage pathway that is encoded by the host chromosome [88]. It has been observed that Tn*5271* was integrated into the BR60 chromosome by the IS*1071*–mediated homologous recombination [91]. Although IS*1071* was first reported to mediate the formation of stable cointegrates of the donor and target molecules in an *E. coli recA* strain, the cointegrates were probably generated not by the function of IS*1071* but by unknown mechanism(s), because the donor and target molecules used in the experiment theoretically cannot be maintained in the same cell. This is consistent with our results that IS*1071* is non-transposable in *E. coli* [131]. On a 85-kb plasmid pCPE3 from *Alcaligenes* sp. CPE3 isolated in Italy is identified a 3-chlorobenzoate-catabolic IS*1071*-composite transposon [23]. Although this transposon is very similar to Tn*5271* found in North America, the two transposons must have been inserted independently into the different plasmids [23]. This suggests that the nearly identical DNA segments containing the *cbaABC* genes are distributed by the combination of IS*1071* with large plasmids.

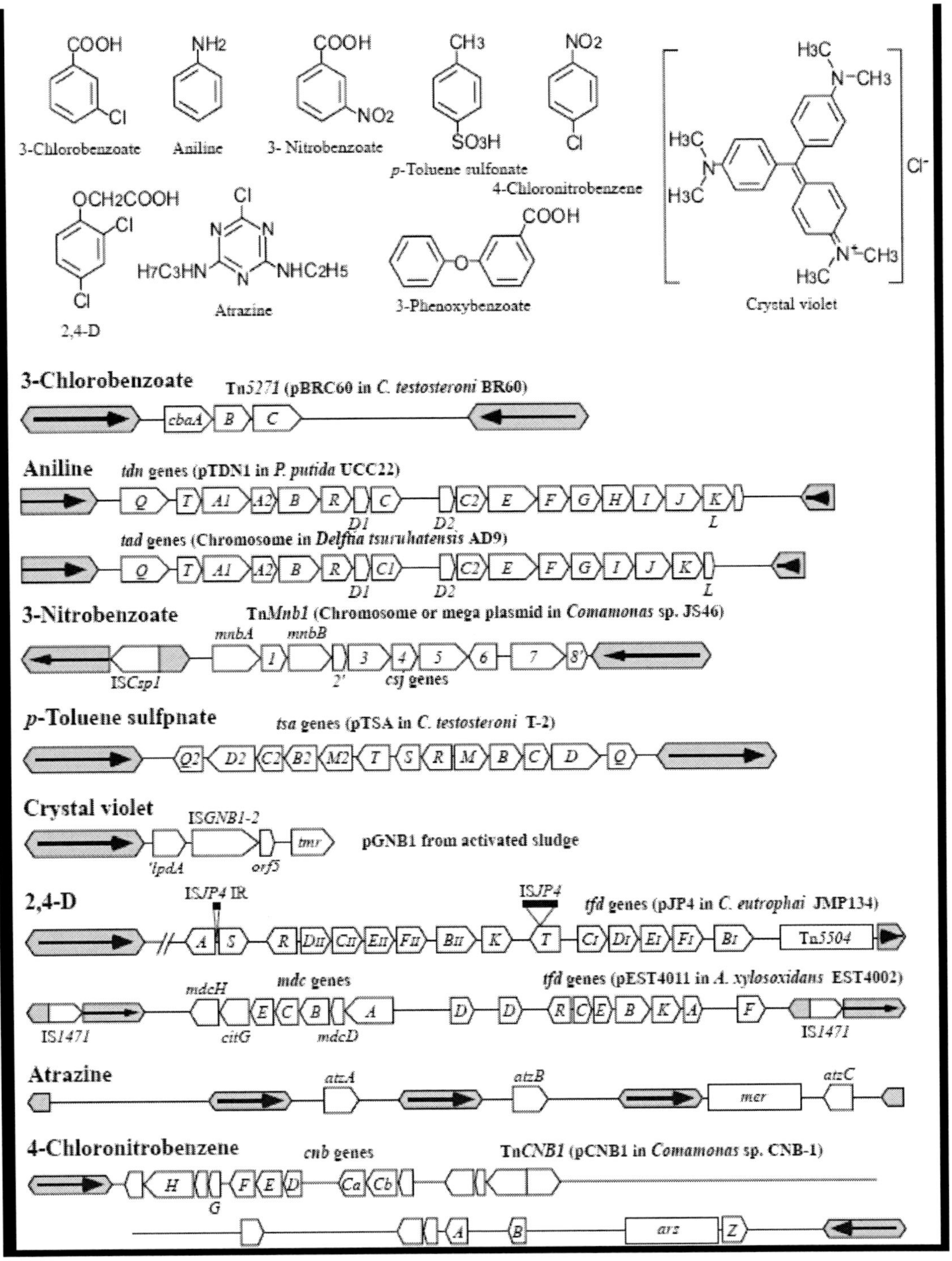

Figure 10. Chemicals and degradation genes associated with IS*1071*. The IS*1071* sequences are depicted by shaded hexagons with arrows indicating the direction. Only the genes involved in the degradation of the target compounds or described in the text are drawn with pentagons. See text for the genes in the figures. The sizes of the figures are arbitrary.

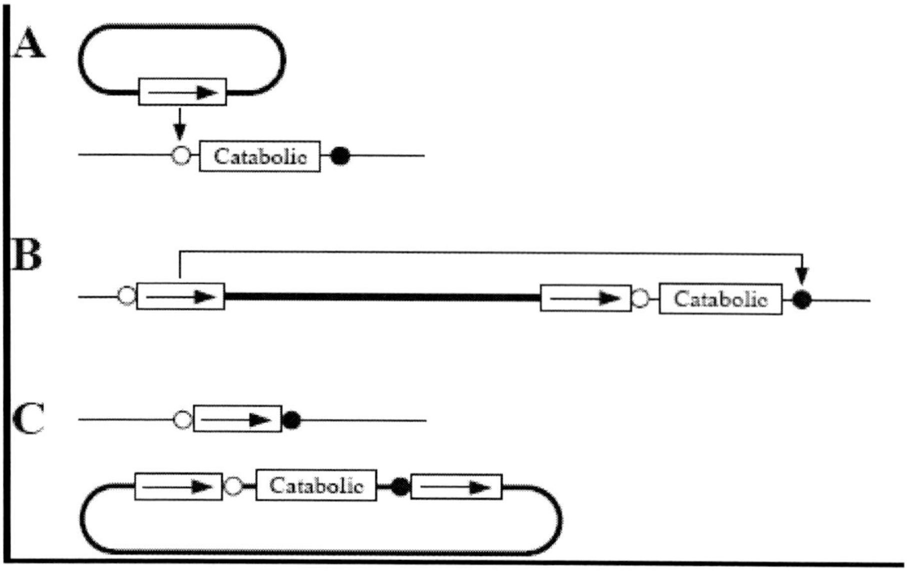

Figure 11. A proposed mechanism for generation of IS*1071*-composite transposons. The figure drawn by Windham et al. [163] is modified. First, IS*1071* (rectangle with an arrow) on a plasmid transposes to a target sequence (white circle) near the catabolic gene (A). The plasmid is integrated into the target molecule by the cointegration function of IS*1071*, and the target sequence is duplicated (B). An intramolecular transposition of an IS*1071* sequence occurs to the novel target sequence (black circle) on the other side of the catabolic gene (folded arrow). The plasmid sequence with the incorporated catabolic gene is excised from the recipient, and the recipient loses the catabolic gene but acquires a copy of IS*1071*. A respective copy of the target duplication generated upon transposition is separated on the plasmid and target molecule.

Aniline Degradation

Aniline is a toxic and recalcitrant aromatic amine compound that is widely used in the manufacture of chemicals such as dyes and herbicides. Plasmid pTDN1 identified in *P. putida* UCC22 confers the host cells to utilize aniline as a growth substrate through the catechol *meta*-cleavage pathway (Figure 3A) [26, 115]. The catabolic genes are located within a 26-kb region bounded by direct repeat sequences, which lead to high-frequency deletion event of the intervening catabolic gene cluster in the absence of aniline [115]. A 9.2-kb segment of pTDN1 was sequenced in 1997, and six genes (*tadQTA1A2BR*) involved in the aniline degradation have been identified (Figure 10). On the basis of sequence similarity to the known genes in databases, the six genes are proposed to encode the large and small subunits of aniline oxygenase (*tdnA1* and *tdnA2*, respectively), a reductase (*tdnB*), and a LysR-type regulator (*tdnR*), and two proteins for amino-group transfer (*tdnQ* and *tdnT*). A 1.8-kb IS*1071* sequence is found upstream of the *tdnQ* gene (Figure 10). Recent sequencing of the 26-kb segment including the 9.2-kb portion has revealed that the *tdnQTA1A2BR* genes (aniline to catechol) and *tdnD1CD2C2EFGHIJKL* genes (catechol *meta*-pathway) are flanked by two copies of IS*1071* [27] (Figure 10). The genes and their organization for the catechol *meta*-pathway show similarity to those for the degradation of other aromatic compounds such as *nah* and *xyl* gene clusters. However, the pTDN1 *meta*-pathway gene cluster has several duplications and insertions when compared with other conserved gene clusters. Thus, the aniline degradation pathway on pTDN1 has probably evolved by the independent acquisition of the aniline upper and catechol *meta*-pathways, the latter of which had undergone several other rearrangements.

Five strains in *Comamonadaceae*, *C. testosteroni* strain I2 and *D. acidovorans* strains CA28, BN3.1, LME1, and B8c, are able to metabolize both aniline and 3-chloroaniline (3-CA) as sole sources of carbon and/or nitrogen [5]. All five strains harbored large plasmids of approximately 100 kb in sizes, and four of them are transferable. The diversity of the plasmids residing in these strains has been examined by the restriction profiles and Southern hybridization. Plasmid pNB2 from strain I2 is responsible for the total degradation of aniline but not for 3-CA degradation, whereas pNB1 from LME1 and pNB8c from B8c catalyze only the oxidative deamination of aniline. Derivatives of pC1 from strain CA28 confer a few patterns of degradability: pC1-1 and pC1-3 enable the host cells to utilize both aniline and 3-CA as nitrogen sources, but pC1-2 does only aniline. The non-transferable plasmid pB1 from BN3.1 allows to use both aniline and 3-CA. All plasmids belong to the IncP-1 group of plasmids based on the PCR amplicons targeted for the IncP-1 *korA* gene. All plasmids except pB1 show several common restriction fragments, and possess the IS*1071* sequence on the 1.4-kb restriction fragments that hybridized with the *tdnQ* probe. Comparison of the partial sequences of the *tdnQ* genes on five plasmids has shown the diversity of the genes with 80 to 98% reciprocal identities. Although it is not sure at present whether the aniline degradation genes on these five plasmids are components of IS*1071*-composite transposons, the observations described above suggest that IS*1071* and IncP-1 plasmids are involved in the dissemination of aniline degradation genes in the environmental bacteria.

The aniline-degradation genes associated with IS*1071* are also identified in another bacterial chromosome. *D. tsuruhatensis* AD9 isolated from the soil surrounding a textile dyeing plant in China carries the whole gene set to convert aniline to the TCA-cycle intermediates [71]. The 25-kb segment responsible for the degradation contains 17 *tad* genes (*tadQTA1A2BRD1D2C2EFGIJKL*) that are bracketed by IS*1071* (Figure 10). The overall structure and nucleotide sequence of this gene cluster are significantly similar to those of pTDN1 from *P. putida* UCC22 [26] except the absence of a gene corresponding to *tdnH* (Figure 10), indicating the distribution of the aniline-degrading IS*1071*-composite transposon between bacterial strains as well as DNA replicons.

3-Nitrobenzoate Degradation

Bacterial aerobic degradation of nitrobenzoate (NBA) isomers has been reported [41, 86, 164]. Whereas the nitro group in 2- or 4-NBA is subjected to reduction preceding to final ring fission, the nitro-group in 3-NBA is spontaneously released from the molecule after two oxygen atoms are added to the 3-, and 4-positions of the aromatic ring, leading to the production of protocatechuate. *Comamonas* sp. JS46 isolated from an industrial wastewater treatment facility in the US can grow on 3-NBA as the source of carbon and nitrogen, but not on 2- or 4-NBA [86]. The strain also utilizes benzoate and 3- and 4-hydroxybenzoates, but not 3-chlorobenzoate, toluene, benzene, phenol, or chlorobenzene. The protocatechuate produced from 3-NBA is then metabolized to 2-hydroxy-4-carboxy-muconic semialdehyde (*meta*-cleavage), which is probably generated by protocatechuate 4,5-dioxygenase [86]. The *mnbA* gene responsible for the 3-NBA conversion to protocatechuate is likely to be carried on the JS46 chromosome or a megaplasmid that is not yet detected. At least seven copies of IS*1071* are identified in JS46, and a spontaneous mutant of JS46 losing the ability to grow on 3-NBA is missing a 10-kb fragment together with one copy of IS*1071*. This loss is assumed to be due to homologous recombination between two copies of IS*1071* that flank the *mnbA* gene, and this assumption has been confirmed by sequence analysis of the corresponding region. The *mnbA* gene is, together with the *mnbB* gene encoding a NAD(P)H-dependent oxidoreductase, carried on an IS*1071*-composite transposon, although one IS*1071* copy has an insertion of another 1.4-kb IS element designated IS*Csp1* (Figure 10). The 17-kb transposon, named as Tn*Mnb1*, also containes eight (six intact and two truncated) genes, the functions of whose products remain to be investigated. Tn*Mnb1* has five alternative structures generated by divergent inversion events, which appeare to be the results of homologous recombination between IRs of IS*1071*. Such rearrangements in an IS*1071*-composite transposon might have affected the diversification and evolution of IS*1071*-composite transposons.

Atrazine Degradation

Degradation genes for atrazine, a herbicide with alternate carbon and nitrogen atoms in the hexagon ring structure (Figure 10), are one of remarkable genetic materials that are closely located to IS*1071*. Many atrazine-degrading bacterial strains, regardless of the G^+ or G^- bacteria, have the IS*1071* sequence on their genomes, particularly on plasmids [22, 77, 112, 113, 144]. Despite that many bacteria have been found to carry the atrazine degradation genes, their detailed sequence determination has rarely been performed. The archetype of such genetic studies on atrazine degradation is that associated with *Pseudomonas* sp. ADP, which was isolated from a herbicide spill site [75]. This strain metabolizes atrazine as a sole nitrogen source. Atrazine is proposed to be metabolized to hydroxyatrazine, cyanuric acid, biuret, urea, and finally carbon dioxide and ammonia [75, 77]. Conversion of atrazine to cyanuic acid is processed via three enzymatic steps mediated by the *atzABC* gene products [8, 18, 114]. The *atzABC* genes are carried on the 109-kb plasmid pADP-1 that belongs to the IncP-1β group [77]. Plasmid pADP-1 possesses between the *tra* and *trb* operons (Figure 9) the 52-kb fragment that contains the *atzABC* genes and five copies of IS*1071* (Figure 10). The two terminal IS*1071* elements are truncated (Figure 10). The three *atz* genes are separated more than 12 kb with one another. Such separate localization of the *atzABC* genes are common in most of bacterial strains carrying the pADP-1-type *atzABC* genes: the *atzA* gene is usually located on a plasmid different from those carrying *atzB* and/or *atzC* genes, and *atzB* and *atzC* genes are sometimes located on different plasmids. The 52-kb fragment on pADP-1 also contains an entire set of mercury-resistance genes similar to those of Tn*5053* [61], several copies of IS*801*, and genes for metabolize dihydrolipoamide. Interestingly, four of the five IS*1071* copies are flanked with the ATGGC target sequence: two copies of such a sequence bracket the two IS*1071* copies, and one such sequence is located just outside of the other two IS*1071* copies. Since IS*1071* does not leave the 5-bp target duplication at both outsides of the element after transposition (see above) [131, 163], these duplications might have been generated by repeated DNA rearrangements such as tandem duplication and subsequent deletion between several copies of IS*1071*.

The remaining three atrazine degradation genes, *atzDEF*, are also carried on pADP-1, but they are located far away from the *atzABC* genes. The *atzDEF* genes are located the *oriV-trfA* region (Figure 9) of pADP-1, one of two hot spots for the insertion of foreign DNA fragments in IncP-1 plasmids [121, 133]. In contrast to the separate distribution of the *atzABC* genes, the *atzDEF* genes form an operon. The AtzD protein shows 58% amino-acid identity to the cyanuric acid amidohydrase (TrzD), and hydrolyzes cyanuric acid to biuret. The latter compound is sequentially metabolized by the AtzE and AtzF proteins, biuret hydrolase and allophanate hydrolase, respectively. Just upstream of the *atzDEF* operon is a LysR-type transcriptional regulator gene that is transcribed divergently from the operon. This regulator gene and the *atzDEF* cluster have similar G+C contents (59 to 61%) as well as codon usage, suggesting that these four genes are incorporated as the unit. It is not clear whether the LysR-type regulator actually controls the expression of the *atzDEF* genes. No IS*1071*-like sequences are present around *atzDEF*, but a putative transposase gene and an integrase-like gene on the *atzDEF*-carrying fragment are situated near the respective boundary sites between the pADP-1 backbone and *atzDEF*-carrying fragment. These DNA recombinases might have involved in the incorporation of the *atzDEF* genes. Overall, the atrazine degradation pathway has probably been established by gathering the six genes from different sources into a cell or a molecule.

Triphenylmethane Dyes Degradation

The IncP-1β plasmid pGNB1 was isolated from the bacterial community of a municipal wastewater treatment plant using an *E. coli* strain as the host for transformation [120]. This 60-kb plasmid encodes degradation of a triphenylmethan dye, crystal violet (Figure 10), resulting in its decolorization. Two accessory regions of pGNB1, one located downstream of the *trfA* replication gene and the other between *tra* and *trb* operons, have been completely sequenced. In the former region is a cryptic class II transposon Tn*5501* that carries the transposase (*tnpA*) and resolvase (*tnpR*) genes, a 1047-bp IS element IS*GNB1-1*, and two other genes (*orf1* and *orf2*). The latter region has the triphenylmethan reductase gene (*tmr*) and a truncated dihydrolipoamide dehydrogenase gene (*'lpdA*) (Figure 10). This truncation is due to insertion of IS*1071* at the 5'-end of the wild-type *lpdA* gene that is found on the atrazine-degradative plasmids pADP-1 [77] and pAA1 [116] from *Pseudomonas* sp. ADP and *Arthrobacter aurescens*, respectively. Both plasmids are known to carry IS*1071* so that this transposon is probably involved in the spread of the *lpdA* gene. The *tmr* gene is situated downstream of the *'lpdA*-IS*GNB1-2*-*orf5* cluster, and no IS*1071* sequences are found downstream of the *tmr* gene (Figure 10). Therefore, it is unclear how the *tmr* gene was incorporated into the ancestor of pGNB1. The *tmr* gene allows the host cells to degrade crystal violet, malachite green, and basic fuchsin. The *tmr* gene is also found from four *D. acidovorans* isolates that are able to decolorize crystal violet. Since IncP-1β plasmids are often found from *D. acidovorans*, one of the four isolates might be the original host of pGNB1.

2,4-Dichlorophenoxyacetic Acid Degradation

A xenobiotic herbicide, 2,4-dichlorophenoxyacetic acid (2,4-D), has been used worldwide for more than a half century. Various soil bacterial strains have been reported that can utilize 2,4-D as a sole carbon source [13, 43, 105, 142, 156]. Microbial degradation of 2,4-D is well studied in its metabolic pathway as well as the associated gene products. The best-characterized 2,4-D degradation genes are the *tfd* genes on the IncP-1β plasmid pJP4 from *Cupriavidus* (formarly *Ralstonia*) *eutropha* JMP134. This strain also degrades 3-chlorobenzoate (3-CB). Both 2,4-D and 3-CB are decomposed by the *tfd* gene products, and the detailed degradation pathway of 2,4-D is described elsewhere [98]. Two chlorocatechol-degrading gene clusters, *tfd-I* and *tfd-II*, are required for efficient degradation of 2,4-D and 3-CB [67, 102, 103]. The *tfd-I* and *tfd-II* clusters contain $tfdB_IF_IE_ID_IC_I$ and $tfdB_{II}F_{II}E_{II}C_{II}D_{II}$ genes, respectively (Figure 10), which encode 2,4-dichlorophenol hydroxylase (*tfdB*), maleylacetate reductase (*tfdF*), dienelactone hydrolase (*tfdE*), chloromuconate cycloisomerase (*tfdD*), and chlorocatechol-1,2-dioxygenase (*tfdC*). The amino-acid identities between the isofunctional Tfd proteins are 62, 60, 38, and 46% for TfdB, TfdC, TfdD, and TfdF, respectively. However, the TfdEI and TfdEII proteins share little identity in their amino-acid sequences. While the $tfdC_ID_IE_I$ products showed the 2- to 50-fold higher activities to the corresponding substrates than the $tfdC_{II}D_{II}E_{II}$ products, the $tfdF_I$ product exhibited a 50-fold lower activity of maleylacetate reductase than the $tfdF_{II}$ product [102]. The remaining *tfd* genes on pJP4 are *tfdA* for 2,4-D/α-ketoglutarate dioxygenase, *tfdR*, *tfdS*, and *tfdT* for LysR-type transcriptional regulators. The *tfdT* gene is truncated by IS*JP4*, two copies of which bracket the *tfd-II* module (Figure 10). The two *tfd* modules, additional catabolic genes, and a cryptic class II transposon Tn*5504* are located within the two copies of IS*1071* element, one of which is truncated by the insertion of Tn*5504* [146] (Figure 10). Attempts were not successful to mobilize the huge (42-kb) IS*1071*-composite transposon with the entire *tfd* genes and the IS*JP4*-composite transposon with the *tfd-II* module, concluding that the two transposons in pJP4 are inactive [146].

Plasmid pIJB1 from *Burkholderia cepacia* 2a is also an IncP-1β plasmid that carries the whole set for the 2,4-D catabolic genes [105]. The 2,4-D dissimilatory genes (*tfdDRCEBKAF*) and malonate degradation genes (*mdcACDEGBH*) are on a 41-kb composite transposon, Tn*5530*, which is terminated by the two copies of IS*1071*::IS*1471* sequence. Since (i) the gene order of the pIJB1 *tfd* cluster is different from those of the *tfd-I* and *tfd-II* clusters in pJP4 and (ii) the *tfd* gene products of pIJB1 and pJP4 show a wide range of amino acid identities (38 to 91%) [105], the three *tfd* clusters are likely to have evolved independently.

Achromobacter xylosoxidans subsp. *denitrificans* EST4002 harbors an IncP-1 plasmid, pEST4011, which also carries the IS*1071*::IS*1471*-flanked 2,4-D degradation genes [156] (Figure 10). This plasmid is the only one IncP-1 plasmid that belongs to the δ subgroup. Since (i) the organizations of the 2,4-D degradation clusters are very similar between pEST4011 and pIJB4, and (ii) pEST4011 also has the malonate catabolic cluster (*mdcADBCEH*), the catabolic clusters on the two plasmids might have derived from the common ancestor. On the other hand, the remaining regions of pEST4011 and pJP4 have little sequence similarity suggesting that they are derived from different ancestor. Due to the insertion of IS*1471* into IS*1071*, the 48-kb composite transposon will not transpose by the IS*1071* transposase. Instead, IS*1471* is apparently intact so that the 2,4-D catabolic genes are theoretically possible to transpose to other replicons.

D. acidovorans P4a carries the 2,4-D degradation genes on its chromosome [43]. Sequence analysis of the cosmid clones of the P4a chromosome has revealed that a 28.4-kb region with the *tfd* genes is sandwiched by IS*1071*. This region contains two 2,4-D catabolic gene clusters, *tfdRCDEF* and *tfdRC*$_{II}$*E*$_{II}$*BKA*. An identical 8-kb sequence containing IS*1071*, IS*1380*, *orfII*, *tfdR* and *tfdC* is occupied invertedly at both ends of the 28.4-kb region, indicating that this IS*1071*-composite transposon might have been formed by duplication of the 8-kb region. The *tfdRCDEF* cluster is nearly identical (>97%) to those of pENH91 (from *C. eutropha* NH9 [96, 97]), pP51 (from *Pseudomonas* sp. P51 [155]), and *P. chlororaphis* RW71 [106, 107], in all of which no IS*1071* sequences are so far identified. By contrast, the *tfdE*$_{II}$*BKA* genes are nearly identical (>98%) to those of pIJB1 [105], pEST4011 [156], and pTV1 (from *Variovorax paradoxus* TV1 [154]), all of which carries the *tfdRCEBKA* cluster. These observations suggest that the pENH91-like *tfdRCDEF* cluster and the pIJB1-like *tfdRCEBKA* cluster have been combined together on a DNA molecule, and then, a homologous recombination occurred between the two *tfdC* genes to give rise to the present structure. No tests have been performed to verify the transposition of the 2,4-D catabolic transposon of P4a.

Carboxydiphenyl Ethers Degradation

P. pseudoalcaligenes POB310 degrades 3- and 4-carboxydiphenyl ethers (3- and 4-CDPE) [19]. These compounds, e.g., 3-phenoxybenzoate (Figure 10), are initially degraded by a dioxygenase that mediates an angular dioxygenation, yielding protocatechuate and phenol as the metabolic intermediates. The strain has two self-transferable plasmids with 40 kb in sizes, and loss of one plasmid, designated pPOB, results in the defect of 4-CDPE metabolism. The genes for the two-component oxygenase, the oxygenase subunit gene *pobA* and a reductase subunit gene *pobB*, are identified on a 4.5-kb fragment of pPOB. The 4.5-kb fragment has been sequenced, and it is revealed that IS*1071* is located upstream of the *pobA* gene that precedes the *pobB* gene. No sequence information is available at present for the other regions of pPOB so that it is uncertain whether another copy of IS*1071* and other genes involved in CDPE degradation are carried on the plasmid. Plasmid pPOB is of particular interest in that it is the smallest catabolic plasmid with self-transferability.

p-Toluenesulfonate Degradation

p-Toluenesulfonate (*p*-TSA) is rarely present in nature but is synthesized in a large amount for the industrial use. Bacterial aerobic degradation of this compound is mediated through 2,3-dioxygenation [25], 1,2-dioxygenation [51], or methyl monooxygenation [15]. Plasmid pTSA identified in *C. testosteroni* T-2 codes for enzymes for the degradation of *p*-TSA through methyl monooxygenation [52, 145]. The TSA monooxygenase (TsaMB) converts *p*-TSA to *p*-sulfobenzylalcohol, the *p*-sulfobenzylalcohol dehydrogenase (TsaC) alters the alcohol group (-CH2OH) to an aldehyde group (-CHO), and the *p*-sulfobenzaldehyde thus produced is changed to *p*-sulfobenzoate by the *p*-sulfobenzaldehyde dehydrogenase (TsaD). The plasmid belongs to the IncP-1β group, and approximately half of the plasmid sequence has been published [52]. A *mer* operon is located between the *oriV* and *trfA* genes of the plasmid, and the genes for *p*-TSA degradation (*tsaMBCD*) are located downstream of the *trb* operon as the components of an IS*1071*-composite transposon. There are two units of the *tsa* genes: the $tsaR\text{-}tsaM_1B_1C_1D_1\text{-}tsaQ_1$ ($tsaMBCD_1$ operon) and $tsaT\text{-}tsaM_2B_2C_2D_2\text{-}tsaQ_2$ ($tsaMBCD_2$ operon) clusters. The two *tsaMBCD-tsaQ* regions are almost identical except for the initial 410-bp sequences and three-nucleotide differences in the *trbB* genes. The RT-PCR analysis has shown that the $tsaMBCD_2$ operon is not expressed [145]. The presence of this silent *tsa* operon is assumed to allow the evolution of enzymes and development of novel pathways in the future. Among sixteen *p*-TSA degraders isolated in four continents, seven have the *tsa* genes that are responsible for the pTSA-like degradation pathway. Moreover, six of the seven isolates carry the *p*-TSA-degrading IS*1071*-composite transposon. The remaining one isolated in Taiwan is losing only the *tsaT* gene. Although three of the seven do not possess any detectable plasmids, these observations suggest that the pTSA-like *p*-TSA degradation pathway has been distributed worldwide by HGT.

4-Chloronitrobenzene Degradation

The 91-kb IncP-1β plasmid pCNB1 from *Comamonas* sp. CNB-1 is responsible for the degradation of 4-chloronitrobenzene (4-CNB), one of the relatively new anthropogenic compounds used for production of drugs, dyes, pesticides, etc. [72]. The plasmid also carries genes for resistance to chromate in a Tn*21*-like class II transposon that is situated between *trfA* and *oriV* [133]. The genes for the 4-CNB degradation (*cnb*) are loaded on 44.5-kb transposon Tn*CNB1*, in which two invertedly oriented and non-identical copies of iso-IS*1071* bracket the *cnb* genes as well as the genes for resistance to arsenate (Figure 10). Curiously, the transposases of the two iso-IS*1071* sequences show only 81% amino acid identity, which has not been observed in other IS*1071*-composite transposons so far identified. This indicates the divergent evolutionary history of the two elements. 4-CNB is metabolized by the following enzymes: chloronitrobenezene nitroreductase (CnbA), hydroxyaminobenzene mutase (CnbB), 5-chloro/2-aminophenol 1,6-dioxygenase (CnbCaCb), aminomuconic semialdehyde dehydrogenase (CnbD), aminomuconate deaminase (CnbH), 4-oxalocrotonate tautomerase (CnbG), 4-oxalocrotonate decarboxylase (CnbF), and 2-oxo-4-pentanoate hydratase (CnbE). The proposed metabolic pathway of 4-CNB is shown in the reference by Ma et al. [72]. The metabolic pathway and its responsible genes show significant similarity to those for nitrobenzene degradation (*nbz*) of *P. putida* HS12 [101]. However, the organizations of the *cnb* and *nbz* genes are different. Whereas all the *cnb* genes are carried on pCNB1, the *nbz* genes are on two different plasmids; *nbzB* (corresponding to *cnbB*) on pNB2 and the others on pNB1. Although the *nbzCaCbDGFEIH* genes are consecutive in pNB1, the corresponding *cnb* genes are split into several clusters by insertion of transposons (Figure 10). Moreover, the nucleotide identities between corresponding *cnb* and *nbz* genes vary 40 to 92%. These observations suggest that the *cnb* and *nbz* genes have evolved and assembled independently. From the viewpoint of genetic and enzymatic evolution, it is of interest to find what changes in amino acid sequences of the Nbz proteins have allowed the proteins to degrade 4-CNB, leading to the generation of Cnb proteins.

Conclusion

We have reviewed the class II and IS*1071*-related catabolic transposons known to date. The information described above strongly suggests that these elements have contributed greatly to the dissemination of catabolic gene clusters and to the creation of novel catabolic pathways in bacterial genomes. We have also illustrated that these elements are the key factors for the rapid generation and evolution of catabolic plasmids. However, there are two crucial problems making these elucidations less comprehensible. Firstly, only a small number of catabolic class II transposons have been so far identified. Although the number of the sequenced bacterial genome is increasing exponentially, the number of class II catabolic transposons is not increasing. Secondly, all of the IS*1071*-composite transposons as well as most of the class I IS-composite transposons, have yet not exhibited transposition, although the transposition of the component IS elements themselves has been confirmed. These problems must be solved in the near future to better understand the evolution of bacterial catabolic activities and to use them effectively for the removal of environmental pollutants. Equally, more information on catabolic plasmids with respect to their entire structures and potential hosts is necessary to comprehend the ecology of catabolic plasmids in bacteria. The success of plasmid- and/or transposon-mediated degradation of xenobiotics will largely depend on increased knowledge of these mobile genetic elements.

Acknowledgments

M.S. was supported by NIH Grant # P20 RR16448 from the COBRE Program of the National Center for Research Resources. M.T. was supported by Grant-in-Aids from the Ministry of Education, Culture, Sports, Science and Technology, and the Ministry of Agriculture, Forestry, and Fisheries (HC-07-2323), Japan.

References

[1] Arcos, J. C., Argus, M. F. (1968). Molecular geometry and carcinogenic activity of aromatic compounds. New perspectives. *Advances in Cancer Research, 11*, 305-471.

[2] Arthur, M., Molinas, C., Depardieu, F., Courvalin, P. (1993). Characterization of Tn*1546*, a Tn*3*-related transposon conferring glycopeptide resistance by synthesis of

depsipeptide peptidoglycan precursors in *Enterococcus faecium* BM4147. *Journal of Bacteriology*, *175*, 117-127.

[3] Assinder, S. J. et al. (1993). A comparison of the multiple alleles of *xylS* carried by TOL plasmids pWW53 and pDK1 and its implications for their evolutionary relationship. *Journal of General Microbiology*, *139*, 557-568.

[4] Benson, S., Shapiro, J. (1978). TOL is a broad-host-range plasmid. *Journal of Bacteriology*, *135*, 278-280.

[5] Boon, N., Goris, J., De Vos, P., Verstraete, W., Top, E. M. (2001). Genetic diversity among 3-chloroaniline- and aniline-degrading strains of the *Comamonadaceae*. *Applied and Environmental Microbiology*, *67*, 1107-1115.

[6] Bosch, R., Garcia-Valdes, E., Moore, E. R. (1999). Genetic characterization and evolutionary implications of a chromosomally encoded naphthalene-degradation upper pathway from *Pseudomonas stutzeri* AN10. *Gene*, *236*, 149-157.

[7] Bosch, R., Garcia-Valdes, E., Moore, E. R. (2000). Complete nucleotide sequence and evolutionary significance of a chromosomally encoded naphthalene-degradation lower pathway from *Pseudomonas stutzeri* AN10. *Gene*, *245*, 65-74.

[8] Boundy-Mills, K. L., de Souza, M. L., Mandelbaum, R. T., Wackett, L. P., Sadowsky, M. J. (1997). The *atzB* gene of *Pseudomonas* sp. strain ADP encodes the second enzyme of a novel atrazine degradation pathway. *Applied and Environmental Microbiology*, *63*, 916-923.

[9] Burlage, R. S., Bemis, L. A., Layton, A. C., Sayler, G. S., Larimer, F. (1990). Comparative genetic organization of incompatibility group P degradative plasmids. *Journal of Bacteriology*, *172*, 6818-6825.

[10] Burrus, V., Pavlovic, G., Decaris, B., Guedon, G. (2002). Conjugative transposons: the tip of the iceberg. *Molecular Microbiology*, *46*, 601-610.

[11] Chain, P. S. et al. (2006). *Burkholderia xenovorans* LB400 harbors a multi-replicon, 9.73-Mbp genome shaped for versatility. *Proceedings of the National Academy of Sciences of the United States of America*, *103*, 15280-15287.

[12] Chandler, M., Mahillon, J. (2002). Insertion sequences revisited. In: N. L. Craig, R. Craigie, M. Gellert, A. M. Lambowitz (eds), *Mobile DNA II*, pp. 305-366). Washington, D. C.: ASM Press.

[13] Chaudhry, G. R., Huang, G. H. (1988). Isolation and characterization of a new plasmid from a *Flavobacterium* sp. which carries the genes for degradation of 2,4-dichlorophenoxyacetate. *Journal of Bacteriology*, *170*, 3897-3902.

[14] Chiou, C. S., Jones, A. L. (1993). Nucleotide sequence analysis of a transposon (Tn*5393*) carrying streptomycin resistance genes in *Erwinia amylovora* and other gram-negative bacteria. *Journal of Bacteriology*, *175*, 732-740.

[15] Cook, A. M., Laue, H., Junker, F. (1998). Microbial desulfonation. *FEMS Microbiology Reviews*, *22*, 399-419.

[16] Dahlberg, C., Hermansson, M. (1995). Abundance of Tn*3*, Tn*21*, and Tn*501* transposase (tnpA) sequences in bacterial community DNA from marine environments. *Applied and Environmental Microbiology*, *61*, 3051-3056.

[17] de la Cruz, F., Grinsted, J. (1982). Genetic and molecular characterization of Tn*21*, a multiple resistance transposon from R100.1. *Journal of Bacteriology*, *151*, 222-228.

[18] de Souza, M. L., Wackett, L. P., Boundy-Mills, K. L., Mandelbaum, R. T., Sadowsky, M. J. (1995). Cloning, characterization, and expression of a gene region from *Pseudomonas* sp. strain ADP involved in the dechlorination of atrazine. *Applied and Environmental Microbiology*, *61*, 3373-3378.

[19] Dehmel, U., Engesser, K. H., Timmis, K. N., Dwyer, D. F. (1995). Cloning, nucleotide sequence, and expression of the gene encoding a novel dioxygenase involved in metabolism of carboxydiphenyl ethers in *Pseudomonas pseudoalcaligenes* POB310. *Archives of Microbiology*, *163*, 35-41.

[20] Dennis, J. J., Zylstra, G. J. (2004). Complete sequence and genetic organization of pDTG1, the 83 kilobase naphthalene degradation plasmid from *Pseudomonas putida* strain NCIB 9816-4. *Journal of Molecular Biology*, *341*, 753-768.

[21] Dennis, J. J. (2005). The evolution of IncP catabolic plasmids. *Current Opinion in Biotechnology*, *16*, 291-298.

[22] Devers, M., El Azhari, N., Kolic, N. U., Martin-Laurent, F. (2007). Detection and organization of atrazine-degrading genetic potential of seventeen bacterial isolates belonging to divergent taxa indicate a recent common origin of their catabolic functions. *FEMS Microbiology Letters*, *273*, 78-86.

[23] Di Gioia, D., Peel, M., Fava, F., Wyndham, R. C. (1998). Structures of homologous composite transposons carrying *cbaABC* genes from Europe and North America. *Applied and Environmental Microbiology*, *64*, 1940-1946.

[24] Dunn, N. W., Gunsalus, I. C. (1973). Transmissible plasmid coding early enzymes of naphthalene oxidation in *Pseudomonas putida*. *Journal of Bacteriology*, *114*, 974-979.

[25] Focht, D. D., Williams, F. D. (1970). The degradation of *p*-toluenesulfonate by a *Pseudomonas*. *Canadian Journal of Microbiology*, *16*, 309-316.

[26] Fukumori, F., Saint, C. P. (1997). Nucleotide sequences and regulational analysis of genes involved in conversion of aniline to catechol in *Pseudomonas putida* UCC22(pTDN1). *Journal of Bacteriology*, *179*, 399-408.

[27] 27. Fukumori, F., Saint, C. P. (2001). Complete nucleotide sequence of the catechol metabolic region of plasmid pTDN1. *J Gen Appl Microbiol*, *47*, 329-333.

[28] Gallegos, M. T., Williams, P. A., Ramos, J. L. (1997). Transcriptional control of the multiple catabolic pathways encoded on the TOL plasmid pWW53 of *Pseudomonas putida* MT53. *Journal of Bacteriology*, *179*, 5024-5029.

[29] Genka, H., Nagata, Y., Tsuda, M. (2002). Site-specific recombination system encoded by toluene catabolic transposon Tn*4651*. *Journal of Bacteriology*, *184*, 4757-4766.

[30] Gogarten, J. P., Townsend, J. P. (2005). Horizontal gene transfer, genome innovation and evolution. *Nature Reviews of Microbiology*, *3*, 679-687.

[31] Goyal, A. K., Zylstra, G. J. (1997). Genetics of naphthalene and phenanthrene degradation by *Comamonas testosteroni*. *Journal of Industrial Microbiology and Biotechnology*, *19*, 401-407.

[32] Grainge, I., Jayaram, M. (1999). The integrase family of recombinase: organization and function of the active site. *Molecular Microbiology*, *33*, 449-456.

[33] Greated, A., Lambertsen, L., Williams, P. A., Thomas, C. M. (2002). Complete sequence of the IncP-9 TOL plasmid pWW0 from *Pseudomonas putida*. *Environmental Microbiology*, *4*, 856-871.

[34] Grimm, A. C., Harwood, C. S. (1999). NahY, a catabolic plasmid-encoded receptor required for chemotaxis of *Pseudomonas putida* to the aromatic hydrocarbon naphthalene. *Journal of Bacteriology*, *181*, 3310-3316.

[35] Grindley, N. D., Reed, R. R. (1985). Transpositional recombination in prokaryotes. *Annual Review of Biochemistry*, *54*, 863-896.

[36] Grindley, N. D. F. (2002). The movement of Tn*3*-like elements: Transposition and cointegrate resolution. In: N. L. Craig, R. Craigie, M. Gellert, A. M. Lambowitz (eds), *Mobile DNA II*, pp. 272-302). Washington, D. C.: ASM Press.

[37] Grinsted, J., de la Cruz, F., Schmitt, R. (1990). The Tn*21* subgroup of bacterial transposable elements. *Plasmid*, *24*, 163-189.

[38] Harada, K. M., Aso, Y., Hashimoto, W., Mikami, B., Murata, K. (2006). Sequence and analysis of the 46.6-kb plasmid pA1 from *Sphingomonas* sp. A1 that corresponds to the typical IncP-1β plasmid backbone without any accessory gene. *Plasmid*, *56*, 11-23.

[39] Harayama, S., Rekik, M., Wasserfallen, A., Bairoch, A. (1987). Evolutionary relationships between catabolic pathways for aromatics: conservation of gene order and nucleotide sequences of catechol oxidation genes of pWW0 and NAH7 plasmids. *Molecular and General Genetics*, *210*, 241-247.

[40] Harwood, C. S., Parales, R. E. (1996). The β-ketoadipate pathway and the biology of self-identity. *Annual Review of Microbiology*, *50*, 553-590.

[41] Hasegawa, Y., Muraki, T., Tokuyama, T., Iwaki, H., Tatsuno, M., Lau, P. C. (2000). A novel degradative pathway of 2-nitrobenzoate via 3-hydroxyanthranilate in *Pseudomonas fluorescens* strain KU-7. *FEMS Microbiology Letters*, *190*, 185-190.

[42] Heinemann, J. A., Sprague, G. F., Jr. (1989). Bacterial conjugative plasmids mobilize DNA transfer between bacteria and yeast. *Nature*, *340*, 205-209.

[43] Hoffmann, D., Kleinsteuber, S., Müller, R. H., Babel, W. (2003). A transposon encoding the complete 2,4-dichlorophenoxyacetic acid degradation pathway in the alkalitolerant strain *Delftia acidovorans* P4a. *Microbiology*, *149*, 2545-2556.

[44] Horak, R., Kivisaar, M. (1999). Regulation of the transposase of Tn*4652* by the transposon-encoded protein TnpC. *Journal of Bacteriology*, *181*, 6312-6318.

[45] Jacoby, G. A., Rogers, J. E., Jacob, A. E., Hedges, R. W. (1978). Transposition of *Pseudomonas* toluene-degrading genes and expression in *Escherichia coli*. *Nature*, *274*, 179-180.

[46] Jain, R., Rivera, M. C., Moore, J. E., Lake, J. A. (2002). Horizontal gene transfer in microbial genome evolution. *Theoretical Population Biology*, *61*, 489-495.

[47] Jain, R., Rivera, M. C., Moore, J. E., Lake, J. A. (2003). Horizontal gene transfer accelerates genome innovation and evolution. *Molecular Biology and Evolution*, *20*, 1598-1602.

[48] Jeenes, D. J., Williams, P. A. (1982). Excision and integration of degradative pathway genes from TOL plasmid pWW0. *Journal of Bacteriology*, *150*, 188-194.

[49] Jencova, V., Strnad, H., Chodora, Z., Ulbrich, P., Hickey, W. J., Paces, V. (2004). Chlorocatechol catabolic enzymes from *Achromobacter xylosoxidans* A8. *Int. Biodeterior. Biodegradation*, *54*, 175-181.

[50] Jencova, V. et al. (2008). Nucleotide sequence, organization and characterization of the (halo)aromatic acid catabolic plasmid pA81 from *Achromobacter xylosoxidans* A8. *Research in Microbiology, In Press,*

[51] Junker, F., Leisinger, T., Cook, A. M. (1994). 3-Sulphocatechol 2,3-dioxygenase and other dioxygenases (EC 1.13.11.2 and EC 1.14.12.-) in the degradative pathways of 2-aminobenzenesulphonic, benzenesulphonic and 4-toluenesulphonic acids in *Alcaligenes* sp. strain O-1. *Microbiology, 140*, 1713-1722.

[52] Junker, F., Cook, A. M. (1997). Conjugative plasmids and the degradation of arylsulfonates in *Comamonas testosteroni. Applied and Environmental Microbiology, 63*, 2403-2410.

[53] Kado, C. I. (2000). *Horizontal gene transfer* (Second edn). London: Academic Press.

[54] Kamachi, K. et al. (2006). Plasmid pBP136 from *Bordetella pertussis* represents an ancestral form of IncP-1β plasmids without accessory mobile elements. *Microbiology, 152*, 3477-3484.

[55] Kanaly, R. A., Harayama, S. (2000). Biodegradation of high-molecular-weight polycyclic aromatic hydrocarbons by bacteria. *Journal of Bacteriology, 182*, 2059-2067.

[56] Kawasaki, H., Tone, N., Tonomura, K. (1981). Plasmid-determined dehalogenation of haloacetates in *Moraxella* species. *Agricultural and Biological Chemistry, 45*, 29-34.

[57] Kawasaki, H., Yahara, H., Tonomura, K. (1981). Isolation and characterization of plasmid pUO1 mediating dehalogenation of haloacetate and mercury resistance in *Moraxella* sp. B. *Agricultural and Biological Chemistry, 45*, 1477-1481.

[58] Keil, H., Keil, S., Pickup, R. W., Williams, P. A. (1985). Evolutionary conservation of genes coding for *meta* pathway enzymes within TOL plasmids pWW0 and pWW53. *Journal of Bacteriology, 164*, 887-895.

[59] Keil, H., Keil, S., Williams, P. A. (1987). Molecular analysis of regulatory and structural *xyl* genes of the TOL plasmid pWW53-4. *Journal of General Microbiology, 133*, 1149-1158.

[60] Keil, H., Saint, C. M., Williams, P. A. (1987). Gene organization of the first catabolic operon of TOL plasmid pWW53: production of indigo by the *xylA* gene product. *Journal of Bacteriology, 169*, 764-770.

[61] Kholodii, G. Y., Yurieva, O. V., Lomovskaya, O. L., Gorlenko, Z., Mindlin, S. Z., Nikiforov, V. G. (1993). Tn*5053*, a mercury resistance transposon with integron's ends. *Journal of Molecular Biology, 230*, 1103-1107.

[62] Kholodii, G. Y., Mindlin, S. Z., Bass, I. A., Yurieva, O. V., Minakhina, S. V., Nikiforov, V. G. (1995). Four genes, two ends, and a *res* region are involved in transposition of Tn*5053*: a paradigm for a novel family of transposons carrying either a mer operon or an integron. *Molecular Microbiology, 17*, 1189-1200.

[63] Kholodii, G. Y. et al. (1997). Tn*5041*: a chimeric mercury resistance transposon closely related to the toluene degradative transposon Tn*4651*. *Microbiology, 143*, 2549-2556.

[64] Kiyohara, H. et al. (1994). Cloning and characterization of a chromosomal gene cluster, *pah*, that encodes the upper pathway for phenanthrene and naphthalene utilization by *Pseudomonas putida* OUS82. *Journal of Bacteriology, 176*, 2439-2443.

[65] Kurland, C. G., Canback, B., Berg, O. G. (2003). Horizontal gene transfer: a critical view. *Proceedings of the National Academy of Sciences of the United States of America, 100*, 9658-9662.

[66] L'Abee-Lund, T. M., Sorum, H. (2000). Functional Tn*5393*-like transposon in the R plasmid pRAS2 from the fish pathogen *Aeromonas salmonicida* subspecies salmonicida isolated in Norway. *Applied and Environmental Microbiology, 66*, 5533-5535.

[67] Laemmli, C. M., Leveau, J. H., Zehnder, A. J., van der Meer, J. R. (2000). Characterization of a second *tfd* gene cluster for chlorophenol and chlorocatechol metabolism on plasmid pJP4 in *Ralstonia eutropha* JMP134(pJP4). *Journal of Bacteriology, 182*, 4165-4172.

[68] Lehrbach, P. R., McGregor, I., Ward, J. M., Broda, P. (1983). Molecular relationships between pseudomonas Inc P-9 degradative plasmids TOL, NAH, and SAL. *Plasmid, 10*, 164-174.

[69] Lessner, D. J., Parales, R. E., Narayan, S., Gibson, D. T. (2003). Expression of the nitroarene dioxygenase genes in *Comamonas* sp. strain JS765 and *Acidovorax* sp. strain JS42 is induced by multiple aromatic compounds. *Journal of Bacteriology, 185*, 3895-3904.

[70] Li, W. et al. (2004). Complete nucleotide sequence and organization of the naphthalene catabolic plasmid pND6-1 from *Pseudomonas* sp. strain ND6. *Gene, 336*, 231-240.

[71] Liang, Q., Takeo, M., Chen, M., Zhang, W., Xu, Y., Lin, M. (2005). Chromosome-encoded gene cluster for the metabolic pathway that converts aniline to TCA-cycle intermediates in *Delftia tsuruhatensis* AD9. *Microbiology, 151*, 3435-3446.

[72] Ma, Y. F. et al. (2007). Nucleotide sequence of plasmid pCNB1 from *Comamonas* strain CNB-1 reveals novel genetic organization and evolution for 4-chloronitrobenzene degradation. *Applied and Environmental Microbiology, 73*, 4477-4483.

[73] Maeda, K., Nojiri, H., Shintani, M., Yoshida, T., Habe, H., Omori, T. (2003). Complete nucleotide sequence of carbazole/dioxin-degrading plasmid pCAR1 in *Pseudomonas resinovorans* strain CA10 indicates its mosaicity and the presence of large catabolic transposon Tn*4676*. *Journal of Molecular Biology, 326*, 21-33.

[74] Mahillon, J., Lereclus, D. (1988). Structural and functional analysis of Tn*4430*: identification of an integrase-like protein involved in the co-integrate-resolution process. *EMBO Journal, 7*, 1515-1526.

[75] Mandelbaum, R. T., Allan, D. L., Wackett, L. P. (1995). Isolation and characterization of a *Pseudomonas* sp. that mineralizes the *s*-triazine herbicide atrazine. *Applied and Environmental Microbiology, 61*, 1451-1457.

[76] Mantengoli, E., Rossolini, G. M. (2005). Tn*5393*d, a complex Tn*5393* derivative carrying the PER-1 extended-spectrum b-lactamase gene and other resistance determinants. *Antimicrobial Agents and Chemotherapy, 49*, 3289-3296.

[77] Martinez, B., Tomkins, J., Wackett, L. P., Wing, R., Sadowsky, M. J. (2001). Complete nucleotide sequence and organization of the atrazine catabolic plasmid pADP-1 from Pseudomonas sp. strain ADP. *Journal of Bacteriology, 183*, 5684-5697.

[78] Mazel, D. (2006). Integrons: agents of bacterial evolution. *Nature Reviews of Microbiology*, *4*, 608-620.

[79] Menn, F. M., Applegate, B. M., Sayler, G. S. (1993). NAH plasmid-mediated catabolism of anthracene and phenanthrene to naphthoic acids. *Applied and Environmental Microbiology*, *59*, 1938-1942.

[80] Mergeay, M. et al. (2003). *Ralstonia metallidurans*, a bacterium specifically adapted to toxic metals: towards a catalogue of metal-responsive genes. *FEMS Microbiology Reviews*, *27*, 385-410.

[81] Minakhina, S., Kholodii, G., Mindlin, S., Yurieva, O., Nikiforov, V. (1999). Tn*5053* family transposons are *res* site hunters sensing plasmidal *res* sites occupied by cognate resolvases. *Molecular Microbiology*, *33*, 1059-1068.

[82] Mindlin, S. et al. (2001). Mercury resistance transposons of gram-negative environmental bacteria and their classification. *Research in Microbiology*, *152*, 811-822.

[83] Miyazaki, R., Sato, Y., Ito, M., Ohtsubo, Y., Nagata, Y., Tsuda, M. (2006). Complete nucleotide sequence of an exogenously isolated plasmid, pLB1, involved in γ-hexachlorocyclohexane degradation. *Applied and Environmental Microbiology*, *72*, 6923-6933.

[84] Monchy, S. et al. (2007). Plasmids pMOL28 and pMOL30 of *Cupriavidus metallidurans* are specialised in the maximal viable response to heavy metals. *Journal of Bacteriology*,

[85] 85. Mongodin, E. F. et al. (2006). Secrets of soil survival revealed by the genome sequence of *Arthrobacter aurescens* TC1. *PLoS Genet*, *2*, e214.

[86] Nadeau, L. J., Spain, J. C. (1995). Bacterial degradation of *m*-nitrobenzoic acid. *Applied and Environmental Microbiology*, *61*, 840-843.

[87] Nakatsu, C., Ng, J., Singh, R., Straus, N., Wyndham, C. (1991). Chlorobenzoate catabolic transposon Tn*5271* is a composite class I element with flanking class II insertion sequences. *Proceedings of the National Academy of Sciences of the United States of America*, *88*, 8312-8316.

[88] Nakatsu, C. H., Wyndham, R. C. (1993). Cloning and expression of the transposable chlorobenzoate-3,4-dioxygenase genes of *Alcaligenes* sp. strain BR60. *Applied and Environmental Microbiology*, *59*, 3625-3633.

[89] Nakazawa, T., Hayashi, E., Yokota, T., Ebina, Y., Nakazawa, A. (1978). Isolation of TOL and RP4 recombinants by integrative suppression. *Journal of Bacteriology*, *134*, 270-277.

[90] Nash, H. A. (1996). Site-specific recombination: integration, excision, resolution, and inversion of defined DNA segments. In: F. C. Neidhardt (ed), *Escherichia coli and Salmonella* (2nd edn, pp. 2363-2376). Washington, D. C.: ASM Press.

[91] Ng, J., Wyndham, C. (1993). IS*1071*-mediated recombinational equilibrium in *Alcaligenes* sp. BR60 carrying the 3-chlorobenzoate catabolic transposon Tn*5271*. *Canadian Journal of Microbiology*, *39*, 92-100.

[92] Nojiri, H. et al. (1999). Diverse oxygenations catalyzed by carbazole 1,9a-dioxygenase from *Pseudomonas* sp. Strain CA10. *Journal of Bacteriology*, *181*, 3105-3113.

[93] Nojiri, H. et al. (2001). Genetic characterization and evolutionary implications of a *car* gene cluster in the carbazole degrader *Pseudomonas* sp. strain CA10. *Journal of Bacteriology, 183*, 3663-3679.

[94] Nojiri, H., Shintani, M., Omori, T. (2004). Divergence of mobile genetic elements involved in the distribution of xenobiotic-catabolic capacity. *Applied Microbiology and Biotechnology, 64*, 154-174.

[95] Nojiri, H., Sota, M., Shintani, M. (2008). Catabolic plasmids involved in the degradation of polycyclic aromatics. In: E. Schwartz (ed), *Microbial Megaplasmids (Microbiology Monographs)* Heidelberg, Germany: Springer.

[96] Ogawa, N., Miyashita, K. (1995). Recombination of a 3-chlorobenzoate catabolic plasmid from *Alcaligenes eutrophus* NH9 mediated by direct repeat elements. *Applied and Environmental Microbiology, 61*, 3788-3795.

[97] Ogawa, N., Miyashita, K. (1999). The chlorocatechol-catabolic transposon Tn*5707* of *Alcaligenes eutrophus* NH9, carrying a gene cluster highly homologous to that in the 1,2,4-trichlorobenzene-degrading bacterium *Pseudomonas* sp. strain P51, confers the ability to grow on 3-chlorobenzoate. *Applied and Environmental Microbiology, 65*, 724-731.

[98] Ogawa, N., Chakrabarty, A. M., Zaborina, O. (2004). Degradative plasmids. In: G. Phillips, B. Funnel (eds), *Plasmid biology*, pp. 341-376). Washington, D. C.: ASM Press.

[99] Osborne, D. J., Pickup, R. W., Williams, P. A. (1988). The presence of two complete homologous *meta* pathway operons on TOL plasmid pWW53. *Journal of General Microbiology, 134*, 2965-2975.

[100] 1Ouchiyama, N., Zhang, Y., Omori, T., Kodama, T. (1993). Biodegradation of carbazole by *Pseudomonas* spp. CA06 and CA10. *Bioscience, Biotechnology, and Biochemistry, 57*, 455-460.

[101] Park, H. S., Kim, H. S. (2000). Identification and characterization of the nitrobenzene catabolic plasmids pNB1 and pNB2 in *Pseudomonas putida* HS12. *Journal of Bacteriology, 182*, 573-580.

[102] Perez-Pantoja, D., Guzman, L., Manzano, M., Pieper, D. H., Gonzalez, B. (2000). Role of *tfdC(I)D(I)E(I)F(I)* and *tfdD(II)C(II)E(II)F(II)* gene modules in catabolism of 3-chlorobenzoate by *Ralstonia eutropha* JMP134(pJP4). *Applied and Environmental Microbiology, 66*, 1602-1608.

[103] Perez-Pantoja, D., Ledger, T., Pieper, D. H., Gonzalez, B. (2003). Efficient turnover of chlorocatechols is essential for growth of *Ralstonia eutropha* JMP134(pJP4) in 3-chlorobenzoic acid. *Journal of Bacteriology, 185*, 1534-1542.

[104] Poelarends, G. J., Kulakov, L. A., Larkin, M. J., van Hylckama Vlieg, J. E., Janssen, D. B. (2000). Roles of horizontal gene transfer and gene integration in evolution of 1,3-dichloropropene- and 1,2-dibromoethane-degradative pathways. *Journal of Bacteriology, 182*, 2191-2199.

[105] Poh, R., Smith, A., Bruce, I. (2002). Complete characterisation of Tn*5530* from *Burkholderia cepacia* strain 2a (pIJB1) and studies of 2,4-dichlorophenoxyacetate uptake by the organism. *Plasmid, 48*, 1-12.

[106] Potrawfke, T., Timmis, K. N., Wittich, R. M. (1998). Degradation of 1,2,3,4-tetrachlorobenzene by *Pseudomonas chlororaphis* RW71. *Applied and Environmental Microbiology*, *64*, 3798-3806.

[107] Potrawfke, T., Armengaud, J., Wittich, R. M. (2001). Chlorocatechols substituted at positions 4 and 5 are substrates of the broad-spectrum chlorocatechol 1,2-dioxygenase of *Pseudomonas chlororaphis* RW71. *Journal of Bacteriology*, *183*, 997-1011.

[108] Providenti, M. A. et al. (2006). The locus coding for the 3-nitrobenzoate dioxygenase of *Comamonas* sp. strain JS46 is flanked by IS*1071* elements and is subject to deletion and inversion events. *Applied and Environmental Microbiology*, *72*, 2651-2660.

[109] Ramos, J. L., Marques, S., Timmis, K. N. (1997). Transcriptional control of the *Pseudomonas* TOL plasmid catabolic operons is achieved through an interplay of host factors and plasmid-encoded regulators. *Annual Review of Microbiology*, *51*, 341-373.

[110] Reddy, B. R., Shaw, L. E., Sayers, J. R., Williams, P. A. (1994). Two identical copies of IS*1246*, a 1275 base pair sequence related to other bacterial insertion sequences, enclose the *xyl* genes on TOL plasmid pWW0. *Microbiology*, *140*, 2305-2307.

[111] Reineke, W. (1998). Development of hybrid strains for the mineralization of chloroaromatics by patchwork assembly. *Annual Review of Microbiology*, *52*, 287-331.

[112] Rousseaux, S., Hartmann, A., Soulas, G. (2001). Isolation and characterisation of new Gram-negative and Gram-positive atrazine degrading bacteria from different French soils. *FEMS Microbiology Ecology*, *36*, 211-222.

[113] Rousseaux, S., Soulas, G., Hartmann, A. (2002). Plasmid localisation of atrazine-degrading genes in newly described *Chelatobacter* and *Arthrobacter* strains. *FEMS Microbiology Ecology*, *41*, 69-75.

[114] Sadowsky, M. J., Tong, Z., de Souza, M., Wackett, L. P. (1998). AtzC is a new member of the amidohydrolase protein superfamily and is homologous to other atrazine-metabolizing enzymes. *Journal of Bacteriology*, *180*, 152-158.

[115] Saint, C. P., McClure, N. C., Venables, W. A. (1990). Physical map of the aromatic amine and *m*-toluate catabolic plasmid pTDN1 in *Pseudomonas putida*: location of a unique *meta*-cleavage pathway. *Journal of General Microbiology*, *136*, 615-625.

[116] Sajjaphan, K. et al. (2004). *Arthrobacter aurescens* TC1 atrazine catabolism genes *trzN*, *atzB*, and *atzC* are linked on a 160-kilobase region and are functional in *Escherichia coli*. *Applied and Environmental Microbiology*, *70*, 4402-4407.

[117] Sato, S. I., Nam, J. W., Kasuga, K., Nojiri, H., Yamane, H., Omori, T. (1997). Identification and characterization of genes encoding carbazole 1,9a-dioxygenase in *Pseudomonas* sp. strain CA10. *Journal of Bacteriology*, *179*, 4850-4858.

[118] Sato, S. I., Ouchiyama, N., Kimura, T., Nojiri, H., Yamane, H., Omori, T. (1997). Cloning of genes involved in carbazole degradation of *Pseudomonas* sp. strain CA10: nucleotide sequences of genes and characterization of meta-cleavage enzymes and hydrolase. *Journal of Bacteriology*, *179*, 4841-4849.

[119] 119. Schell, M. A., Sukordhaman, M. (1989). Evidence that the transcription activator encoded by the *Pseudomonas putida nahR* gene is evolutionarily related to the transcription activators encoded by the *Rhizobium nodD* genes. *Journal of Bacteriology*, *171*, 1952-1959.

[120] 120. Schlüter, A. et al. (2007). IncP-1β plasmid pGNB1 isolated from a bacterial community from a wastewater treatment plant mediates decolorization of triphenylmethane dyes. *Applied and Environmental Microbiology*, *73*, 6345-6350.

[121] 121. Schlüter, A., Szczepanowski, R., Pühler, A., Top, E. M. (2007). Genomics of IncP-1 antibiotic resistance plasmids isolated from wastewater treatment plants provides evidence for a widely accessible drug resistance gene pool. *FEMS Microbiology Reviews*, *31*, 449-477.

[122] Shapiro, J. A. (1979). Molecular model for the transposition and replication of bacteriophage Mu and other transposable elements. *Proceedings of the National Academy of Sciences of the United States of America*, *76*, 1933-1937.

[123] Sherratt, D. (1989). Tn*3* and related transposable elements: site-specific recombination and transposition. In: D. E. Berg, M. M. Howe (eds), *Mobile DNA*, pp. 163-184). Washington DC: American Society for Microbiology.

[124] Shintani, M., Habe, H., Tsuda, M., Omori, T., Yamane, H., Nojiri, H. (2005). Recipient range of IncP-7 conjugative plasmid pCAR2 from *Pseudomonas putida* HS01 is broader than from other *Pseudomonas* strains. *Biotechnol Lett*, *27*, 1847-1853.

[125] Shintani, M., Yoshida, T., Habe, H., Omori, T., Nojiri, H. (2005). Large plasmid pCAR2 and class II transposon Tn*4676* are functional mobile genetic elements to distribute the carbazole/dioxin-degradative car gene cluster in different bacteria. *Applied Microbiology and Biotechnology*, *67*, 370-382.

[126] Shintani, M. et al. (2006). Characterization of the replication, maintenance, and transfer features of the IncP-7 plasmid pCAR1, which carries genes involved in carbazole and dioxin degradation. *Applied and Environmental Microbiology*, *72*, 3206-3216.

[127] Shintani, M. et al. (2007). The *Sphingomonas* plasmid pCAR3 is involved in complete mineralization of carbazole. *Journal of Bacteriology*, *189*, 2007-2020.

[128] Smith, M. C., Thorpe, H. M. (2002). Diversity in the serine recombinases. *Molecular Microbiology*, *44*, 299-307.

[129] Sota, M., Endo, M., Nitta, K., Kawasaki, H., Tsuda, M. (2002). Characterization of a class II defective transposon carrying two haloacetate dehalogenase genes from *Delftia acidovorans* plasmid pUO1. *Applied and Environmental Microbiology*, *68*, 2307-2315.

[130] Sota, M., Kawasaki, H., Tsuda, M. (2003). Structure of haloacetate-catabolic IncP-1β plasmid pUO1 and genetic mobility of its residing haloacetate-catabolic transposon. *Journal of Bacteriology*, *185*, 6741-6745.

[131] Sota, M. et al. (2006). Functional analysis of unique class II insertion sequence IS*1071*. *Applied and Environmental Microbiology*, *72*, 291-297.

[132] Sota, M. et al. (2006). Genomic and functional analysis of the IncP-9 naphthalene-catabolic plasmid NAH7 and its transposon Tn*4655* suggests catabolic gene spread by a tyrosine recombinase. *Journal of Bacteriology*, *188*, 4057-4067.

[133] Sota, M., Tsuda, M., Yano, H., Suzuki, H., Forney, L. J., Top, E. M. (2007). Region-specific insertion of transposons in combination with selection for high plasmid transferability and stability accounts for the structural similarity of IncP-1 plasmids. *Journal of Bacteriology*, *189*, 3091-3098.

[134] Sota, M., Top, E. M. (2008). Horizontal gene transfer mediated by plasmids. In: G. Lipps (ed), *Plasmids: Current Research and Future Trends* Norwich, UK: Horizon Scientific Press.

[135] Springael, D., Top, E. M. (2004). Horizontal gene transfer and microbial adaptation to xenobiotics: new types of mobile genetic elements and lessons from ecological studies. *Trends in Microbiology*, *12*, 53-58.

[136] Sundin, G. W., Mayfield, C. T., Zhao, Y., Gunasekera, T. S., Foster, G. L., Ullrich, M. S. (2004). Complete nucleotide sequence and analysis of pPSR1 (72,601 bp), a pPT23A-family plasmid from *Pseudomonas syringae* pv. *syringae* A2. *Mol Genet Genomics*, *270*, 462-476.

[137] Takizawa, N. et al. (1994). Identification and characterization of genes encoding polycyclic aromatic hydrocarbon dioxygenase and polycyclic aromatic hydrocarbon dihydrodiol dehydrogenase in *Pseudomonas putida* OUS82. *Journal of Bacteriology*, *176*, 2444-2449.

[138] Tehara, S. K., Keasling, J. D. (2003). Gene Cloning, purification, and characterization of a phosphodiesterase from *Delftia acidovorans*. *Applied and Environmental Microbiology*, *69*, 504-508.

[139] Thomas, C. M., Nielsen, K. M. (2005). Mechanisms of, and barriers to, horizontal gene transfer between bacteria. *Nature Reviews of Microbiology*, *3*, 711-721.

[140] Thorsted, P. B. et al. (1998). Complete sequence of the IncPβ plasmid R751: implications for evolution and organisation of the IncP backbone. *Journal of Molecular Biology*, *282*, 969-990.

[141] Todd, J. D. et al. (2007). Structural and regulatory genes required to make the gas dimethyl sulfide in bacteria. *Science*, *315*, 666-669.

[142] Top, E. M., Holben, W. E., Forney, L. J. (1995). Characterization of diverse 2,4-dichlorophenoxyacetic acid-degradative plasmids isolated from soil by complementation. *Applied and Environmental Microbiology*, *61*, 1691-1698.

[143] Top, E. M., Moenne-Loccoz, Y., Pembroke, T., Thomas, C. M. (2000). Phenotypic traits conferred by plasmids. In: C. M. Thomas (ed), *The Horizontal Gene Pool*, pp. 249-286). Amsterdam: Harwood Academic Publishers.

[144] Topp, E., Zhu, H., Nour, S. M., Houot, S., Lewis, M., Cuppels, D. (2000). Characterization of an atrazine-degrading *Pseudaminobacter* sp. isolated from Canadian and French agricultural soils. *Applied and Environmental Microbiology*, *66*, 2773-2782.

[145] Tralau, T., Cook, A. M., Ruff, J. (2001). Map of the IncP1b plasmid pTSA encoding the widespread genes (*tsa*) for *p*-toluenesulfonate degradation in *Comamonas testosteroni* T-2. *Applied and Environmental Microbiology*, *67*, 1508-1516.

[146] Trefault, N. et al. (2004). Genetic organization of the catabolic plasmid pJP4 from *Ralstonia eutropha* JMP134 (pJP4) reveals mechanisms of adaptation to chloroaromatic pollutants and evolution of specialized chloroaromatic degradation pathways. *Environmental Microbiology*, *6*, 655-668.

[147] Tsuda, M., Iino, T. (1987). Genetic analysis of a transposon carrying toluene degrading genes on a TOL plasmid pWW0. *Molecular and General Genetics*, *210*, 270-276.

[148] Tsuda, M., Iino, T. (1988). Identification and characterization of Tn*4653*, a transposon covering the toluene transposon Tn*4651* on TOL plasmid pWW0. *Molecular and General Genetics*, *213*, 72-77.

[149] Tsuda, M., Minegishi, K., Iino, T. (1989). Toluene transposons Tn*4651* and Tn*4653* are class II transposons. *Journal of Bacteriology*, *171*, 1386-1393.

[150] Tsuda, M., Iino, T. (1990). Naphthalene degrading genes on plasmid NAH7 are on a defective transposon. *Molecular and General Genetics*, *223*, 33-39.

[151] Tsuda, M., Tan, H. M., Nishi, A., Furukawa, K. (1999). Mobile catabolic genes in bacteria. *Journal of Bioscience and Bioengineering*, *87*, 401-410.

[152] Tsuda, M., Genka, H. (2001). Identification and characterization of Tn4656, a novel class II transposon carrying a set of toluene-degrading genes from TOL plasmid pWW53. *Journal of Bacteriology*, *183*, 6215-6224.

[153] Tsuda, M., Sota, M. (2005). Mobile genetic elements for microbial degradation of environmental pollutants (In Japanese). *Protein, Nucleic Acid, Enzyme*, *50*, 1527-1534.

[154] 154. Vallaeys, T., Courde, L., McGowan, C., Wright, A. D., Fulthorpe, R. R. (1999). Phylogenetic analyses indicate independent recruitment of diverse gene cassettes during assemblage of the 2,4-D catabolic pathway. *FEMS Microbiology Ecology*, *28*, 373-382.

[155] van der Meer, J. R., Eggen, R. I., Zehnder, A. J., de Vos, W. M. (1991). Sequence analysis of the *Pseudomonas* sp. strain P51 *tcb* gene cluster, which encodes metabolism of chlorinated catechols: evidence for specialization of catechol 1,2-dioxygenases for chlorinated substrates. *Journal of Bacteriology*, *173*, 2425-2434.

[156] Vedler, E., Vahter, M., Heinaru, A. (2004). The completely sequenced plasmid pEST4011 contains a novel IncP1 backbone and a catabolic transposon harboring *tfd* genes for 2,4-dichlorophenoxyacetic acid degradation. *Journal of Bacteriology*, *186*, 7161-7174.

[157] Voss, J. A., Khedairy, H., Baker, R. F., Benjamin, R. C. (1990). Molecular cloning of the *xylL-xylE* region from the *P. putida* TOL plasmid, pDK1. *SAAS Bulletin, Biochemistry and Biotechnology*, *3*, 54-57.

[158] Ward, J. M., Grinsted, J. (1982). Analysis of the Inc P-1 group plasmids R906 and R751 and their relationship to RP1. *Plasmid*, *8*, 244-252.

[159] White, G. P., Dunn, N. W. (1978). Compatibility and sex specific phage plating characteristics of the TOL and NAH catabolic plasmids. *Genetical Research*, *32*, 207-213.

[160] Williams, P. A., Murray, K. (1974). Metabolism of benzoate and the methylbenzoates by *Pseudomonas putida* (arvilla) mt-2: evidence for the existence of a TOL plasmid. *Journal of Bacteriology*, *120*, 416-423.

[161] Williams, P. A., Jones, R. M., Zylstra, G. (2004). Genomics of catabolic plasmids. In: J.-L. Ramos (ed), *Pseudomonas*, pp. 165-195). New York, U. S. A.: Plenum Publishing Corporation.

[162] Wyndham, R. C., Singh, R. K., Straus, N. A. (1988). Catabolic instability, plasmid gene deletion and recombination in *Alcaligenes* sp. BR60. *Archives of Microbiology*, *150*, 237-243.

[163] Wyndham, R. C., Cashore, A. E., Nakatsu, C. H., Peel, M. C. (1994). Catabolic transposons. *Biodegradation*, *5*, 323-342.

[164] Yabannavar, A. V., Zylstra, G. J. (1995). Cloning and characterization of the genes for *p*-nitrobenzoate degradation from *Pseudomonas pickettii* YH105. *Applied and Environmental Microbiology*, *61*, 4284-4290.

[165] Yano, H. et al. (2007). Complete sequence determination combined with analysis of transposition/site-specific recombination events to explain genetic organization of IncP-7 TOL plasmid pWW53 and related mobile genetic elements. *Journal of Molecular Biology*, *369*, 11-26.

[166] Yen, K. M., Gunsalus, I. C. (1985). Regulation of naphthalene catabolic genes of plasmid NAH7. *Journal of Bacteriology*, *162*, 1008-1013.

[167] Yen, K. M., Serdar, C. M. (1988). Genetics of naphthalene catabolism in pseudomonads. *Critical Reviews in Microbiology*, *15*, 247-268.

[168] Zechner, E. L. et al. (2000). Conjugative-DNA transfer process. In: C. M. Thomas (ed), *The Horizontal Gene Pool*, pp. 87-174). Amsterdam: Harwood Academic Publishers.

[169] Zhou, N. Y., Fuenmayor, S. L., Williams, P. A. (2001). *nag* genes of *Ralstonia* (formerly *Pseudomonas*) sp. strain U2 encoding enzymes for gentisate catabolism. *Journal of Bacteriology*, *183*, 700-708.

In: DNA Transposable Elements Research
Editors: K. Yoshida, M. Aoki, pp. 69-108
ISBN: 978-1-60456-531-7
© 2008 Nova Science Publishers, Inc.

Chapter III

Inversions and Inverted Transpositions as Generators of Universal Genome Properties

Guenter Albrecht-Buehler*
Department of Cell and Molecular Biology, Feinberg School of Medicine
Northwestern University, Chicago, IL 60611, USA

Abstract

Transpositions have been identified as built-in responses to genome shock [17] and as agents of genome variation. In addition, they seem to give rise to several universal genome features that may be essential to distinguish functional genomes from meaningless poly-nucleotide and, thus, may facilitate horizontal genome transfer. Examples of such universals appear to be the universal strand symmetry, sometimes called "Chargaff's second parity rule", and the almost universal tri-nucleotide frequency distribution, called "majority profile". Both could be explained as products of the actions of countless inversions and inverted transpositions in the evolutionary past of the genomes [1-3].

As reported here, these particular products of inversions/transpositions give rise to yet another apparent universal in the following way. The genomes whose triplet profiles are almost identical with the majority profile define a large group of organisms called the 'majority group'. In spite of the fact that the triplet profiles of their genomes are almost identical with the majority profile, they may locally deviate from it, nevertheless. In this chapter I will introduce the graphs of these local deviations as "profile signatures" and show their high degree of conservation during the evolution of orders. In addition, they offer rapid and comprehensive overviews over chromosome rearrangements and areas of synteny.

The signatures themselves are obviously not genome universals. However, the underlying process which causes their local deviations from the majority profile appears to be yet another universal, which can be linked directly to the action of inversions /transpositions.

Introduction

The Distinction between Sense and Nonsense Genomes

How is a genome different from a mere random poly-nucleotide? What if I give the reader 200Mb of a perfectly random-looking sequence of A, C, T, and G. Could (s)he tell whether it is actually some organism's genome? Could a cell? At present our answer rests entirely on the genome's function: The genome sequence must make biological sense. After all, it was derived from the nucleus of a living cell, and so it must be able to support countless biological actions of the cell. In contrast, a random poly-nucleotide that originated from a computer's random number generator or from a primordial soup of randomly mixed nucleotides would clearly not be able to do the same.

This is not an academic question. As long as we cannot tell the difference between a real genome and some gibberish poly-nucleotide by the sequence alone, can we actually claim to be able to read genome sequences? In view of the often huge amounts of non-coding sequences, we cannot rely on gene expression as our guide. Will we be able to navigate huge entire genomes, decipher all the records of their evolutionary past, and identify the encoded controls of replication, development and cell function, if we have little else to go on than their sequence? How can we hope to design artificial genomes that work properly, without also designing 'functional' non-coding parts?

Sequence Universals as Validators of Legitimate Genomes

Granted that we cannot tell at the present time whether a foreign genome makes biological sense, could a cell? I suppose, they should. After all, encounters of a cell with a foreign genome are by no means rare events. They occur for example during fertilization and cell fusion. Maybe, we could learn from them how to do it.

Experimentally, one would need to study this problem by transfecting cytoplasts or cells with either real genome sequences or with synthesized nonsense sequences which are properly enclosed in nuclear envelopes, wrapped in nucleosomes and coiled into a set of chromosomes. Would the experiments show different reactions of the cells, and how soon after the transfection would a difference be manifest?

Obviously, such experiments are extremely difficult to do and to interpret, which is probably the reason that they have not been done. But even if we were able to overcome the technical difficulties, at present we would have to assess the cells' reaction to the intrusion of a foreign genome by the time course of their patterns of transcription and translation. Given the enormous complexity of these patterns, it would be very difficult indeed to identify the parts that relate to the recognition and validation of a foreign genome.

What method could a cell use to validate a foreign genome? It can hardly require the cells to complete the enormous time- and energy consuming task of transcribing and translating it in its entirety. Instead, there should be a limited set of special sequence properties whose presence can tell the cell rapidly that the foreign genome makes biological sense. Whatever these sequence properties may be, I submit that they must be universal because the cells must 'expect' them to be present in every legitimate genome, regardless of species.

As an immediate consequence of their universality, these sequence properties must not be confined to the coding regions. After all, coding regions are predominantly species specific. Therefore, the sequence check must cover also the genome regions that are normally not transcribed, such as introns, pseudo-genes, stretches of repetitive DNA and other non-coding regions.

In short, if we want to assess the legitimacy of genome sequences in the way cells may do it, we need to search for and check the expression of a list of universal features of genome sequences in coding and non-coding regions alike. Examples of two such universals which arbitrary genome sequences do not have and which include all non-coding regions are the universal strand symmetry, sometimes called "Chargaff's second parity rule', and the almost universal tri-nucleotide frequency distribution, called "majority profile" [1-3].[1]

Therefore it would appear that we would be able to assess unknown genomes the better, the more sequence universals we discover and check. If successful, the search for genome universals may even turn up a universal reading and writing "format" of genomes. I suspect that it exists, because nuclei of cells from one organism can be transplanted into cytoplasts from another, and yet the cytoplast is able to read it and become a functional cell. Such universals may also facilitate or even be required for horizontal gene transfer and gene duplications [31] and thus speed up evolution tremendously as the cells from different phyla and kingdoms need not re-invent desirable genes over and over again.

Markers for the Expression of Sequence Universals

It is easy for scientist to check whether a foreign genome sequence is strand symmetrical and whether its triplet profile belongs to the majority class. I believe, it is also a practical and sensitive way to determine in many cases whether an unknown genome is not legitimate.

On the other hand, I am also convinced that cells cannot possibly do the same. According to the definition of these universals, the cells would have to count the numbers of all possible 64 triplets on either strand and compare the resulting distributions with each other and with the majority profile. And even if the cells had methods to count triplets and compare distributions, which they surely do not, the enormous size of many genomes alone would be prohibitive for the task.

However, this does not mean that one should dismiss these sequence universals as mere academic toys for computational biologists. After all, if the cells cannot check the expression of these universals by direct counting, maybe they are be able to check it indirectly. If, for example, there were agents which enforce and maintain the expression of these universals, the cells could simply check whether these agents are present. Finding them in the genome

1 These two universals relate to the tri-nucleotide (codon, triplet) frequency distribution of genome sequences that are larger then 100 kb. The universal strand symmetry states that the numbers of each triplets and its reverse complementary triplet are the same on one and the same strand. The 'almost' universality of the majority profile states that the frequency of each triplet is essentially the same for the majority of species and taxa. They will be described in more detail in the Result section.

could be a sufficiently safe indicator that the genome sequence must express these universals, too. No elaborate counting of huge numbers of triplets would be needed. Is there such a common cause for the strand symmetry and the majority profile?

Transposons as Telltale Elements for Genome Universals

Indeed, the two mentioned genome universals, namely the universal strand symmetry and the majority profile could be explained by the actions of a single agent, namely the countless inversions and inverted transpositions the genome experienced in its evolutionary past [1 - 3]. The first part of the Result section will present the evidence in some detail.

Therefore, one may conjecture that cells only need to check the presence of invertible transposable elements if they want to ensure the strand symmetry and majority profile of a foreign genome.

(Note that the cells do not need to check the presence of *specific* transposable elements. The strand symmetry and majority profile of a genome can be created with any or all transposable elements as long as they permit inversion [1,3].)

Hybrid Dysgenesis as a Possible Paradigm

How then could cells check whether a foreign genome contains transposable elements? It would be quite simple, if there was a class of repressor proteins that could generically inhibit transposable elements. I suspect that such generic repressors exist because all genomes may have needed them many times in their evolutionary past in order to protect their coding regions from the mutagenic effects of transposable elements.

These repressors can be used in the following way to check whether a foreign genome contains sufficiently many transposable elements: After unleashing a barrage of its own transposable elements the cell could allow them to 'attack' the foreign genome (and unfortunately also itself). If the foreign genome is unable to counter the 'attack' by expressing the appropriate repressor genes, both genomes may be damaged, and the cell or its descendents may die. However, if the foreign genome can mount a proper defense, both will survive. Obviously, in this case the foreign genome contains the repressor genes, suggesting that it needed them as defense against its own transposable elements. In other words, the cell can 'infer' that the intruder expresses these genome universals because it contains transposable elements which, in turn, must have enforced the strand symmetry and the majority profile in its past.

This method of genome validation may appear quite risky or even suicidal for a cell. On second thought, it is actually reasonable, because a cell that is invaded by a foreign, incompatible genome may be doomed, anyway. In fact, nature seems to use this method in the special case of 'hybrid dysgenesis' [32].

In this case the foreign genome is the male pro-nucleus of the sperm of the so-called P-strain of *Drosophila melanogaster* that enters the oocyte of a so-called M-strain female during fertilization.[2] The result are sterility and extensive mutation in the offspring, because the M-female genome is damaged in the germ line by the mutagenic attacks by the transposable P-elements of the P-male's genome. They are successful because the M-female's genome cannot defend itself as it lacks the P-elements and, hence, has no repressor protein for their transposase, either.

A Third Genome Universal That Is Also Generated by Inversions/Transpositions ('Genome Signatures' and Their Generator)

In biology, one must always be prepared to find that rules are violated. The concept of genome universals is no exception. Many parts of a genome may locally deviate from such a universal. But this does not mean that the existence of exceptions merely weaken the validity of the concept of genome universals. On the contrary, the deviations may point to important biological functions they have.

Therefore, once a genome universal has been found, every part of a genome that deviates from it requires attention. The functions describing these deviations will be called 'genome signatures'. For example, the function describing a genome's local deviations from the universal strand symmetry will be introduced as its 'strand symmetry signature'. Likewise, the genomes of the so-called majority group whose triplet profiles are almost identical with the majority profile were found to deviate from it in several limited stretches of the genome sequences. In this chapter I will introduce graphic representations of these deviations as 'profile signatures' and show their high degree of conservation during the evolution of orders.

To be sure, the profile signatures themselves are not genome universals. However, the present chapter will show that they deviate locally the more from the majority profile, the lower their local AT-content is. This correlation with the local AT-content of a genome gives rise to the so-called the 'universal sigmoid curve', which will be presented below. It appears to be a genome universal. The present chapter will suggests how it could be yet another product of inversions/transpositions, as well.

---

2 The P-strain is characterized by the presence of transposable P-elements which contain their own transposase. The M-strain has no P-elements. Of course, a cross between an M-male and an M-female involves no actions of P-elements because they are absent. Yet, even a cross between a P-male and a P-female does not involve them, because the eggs of P-females contain a large amount of a repressor protein for the transposase of the P-elements. However, the situation changes dramatically, if a P-male is crossed with an M-female. Even though the development of the zygote is allowed to proceed normally (the somatic cells of the offspring fly cannot splice the P-element transposase properly), the P-elements in the germ line of the offspring fly use their built-in transposase to mutate it so extensively, that the offspring is rendered sterile.

Results

A. Background

The Transposition/Inversion Hypothesis of Chargaff's Second Parity Rule

There is growing evidence of large numbers of Alu, SINE, LINE and other such dispersed, repetitive sequences in the coding and non-coding regions of the genomes of many species [13-16]. They suggest that genomes experienced countless inversions and inverted transpositions during evolutionary times. One of their effects appears to be the validity of a remarkable symmetry, sometimes called Chargaff's second parity rule formulated here in the case of tri-nucleotides (triplets) [4-12]: In genome duplexes that exceed 100 kb the frequency distributions of their tri-nucleotides (triplet profiles) are the same in both strands. There is an even more startling equivalent formulation: In genome duplexes that exceed 100 kb the number of every triplet is equal to the number of its complementary triplet on the *same* strand. Obviously, this symmetry cannot be the result of base pairing, because as a rule the triplets do not pair with their complements on the same strand. With only few exceptions they pair with their complements on the opposite strand. Therefore, an entirely different explanation is needed.

Obviously, this peculiar strand symmetry is a species-independent property of genomes. Its expression in genomes is almost universal even though it offers no known selective advantage. After all, if it is universal then its expression in a particular organism does not seem able to distinguish it from its Darwinian competitors. Worse, there seems to be no function associated with this symmetry. What good would it do an organism if the numbers of every triplet and its complementary triplet are the same on (say) the Plus-strand of its genome? What kind of biological function needs to count that? The question is aggravated by the fact that this kind of symmetry applies to coding and non-coding regions equally well. The only exceptions found so far are the mitochondrial genomes of some of the most recent vertebrates. Why would the expression of strand symmetry be universal even in non-coding regions that seem to have no function, and why would many mitochondrial genomes that are so vital to organisms be allowed to violate the rule?

These questions are rather disturbing because we are used to thinking that the expression of all wide spread properties among organisms must be the result of the selective advantages of these properties. If there are no such advantages the many mechanisms of genome variation should long have randomized the expression of these properties at will without any survival threatening consequences for the organisms. It seems that the solution to this dilemma is the answer that the strand symmetry itself may have no fundamental function, but it is the inevitable consequence of something that has.

How then can one explain this strand symmetry? A number of explanations have been put forward in the past [4, 5, 11, 12]. In agreement with the authors of earlier hypotheses about the rule, I assume that all genomes initially violated it because they contained arbitrary numbers of single nucleotides and triplets. Only their subsequent evolution rendered them increasingly compliant with Chargaff's second parity rule by a number of different mechanisms. However, consistent with the topic of this volume, I will focus on the possibility to explain it as the inevitable and asymptotically more and more accurate result of countless inversions /transpositions in the evolutionary past of genomes [4, 5, 1].

To be sure, reversing the order of bases of a transposon would not be able to accomplish the strand symmetry. It would simply add its specific content of triplets to the overall triplet spectrum of the genome. The reason that the inversion of transposable elements, transposons, or other inserted segments has a significant equalizing effect on the strands is the necessity for the inserted elements to swap strands upon inversion. Without swapping strands the 5' ends of the inverted segment would not be able to find 3' ends with which to link up. In other words, the particular strand of such an inversion that was part of a Plus-strand before its excision has to be inserted into the Minus-strand and vice versa. As shown below, this action must equalize in an asymptotic fashion the base composition and oligo-nucleotide composition of the genome in question [1].

Of course, the individual steps involved in the actual mechanism of inversions and transpositions [17] are much more complex. For example, retroposons such as L1 elements involve even RNA intermediates [18]. Yet, these complexities can be ignored in the present case because only net changes of the genome sequence matter for present consideration. Among these, however, the chapter will ignore small direct repeats and other short variations generated by inversions and inverted transpositions. They are usually less than 10 bases long and thus contribute very little to the overall nucleotide statistics compared to the often many 100 kb large inversions and 1-3 kb and larger transposons [13, 14].

Qualitative Description of the Transposition/Inversion Model of Strand Symmetry

Assume that initially the number of G's is much larger than the number of C's on a Plus-strand. Therefore, due to base pairing the Minus-strand contains the opposite imbalance, namely more C's than G's. Due to its strand swapping effect, every randomly located transposition/inversion must carry some of the supernumerary G's from the Plus-strand to the Minus-strand while, at the same time, it carries some of the supernumerary C's from the Minus-strand to the Plus-strand. The result is an ongoing equalization of the numbers of G's with C's on both strands. In a similar way, the mechanism equalizes the numbers of A's and T's on each strand. In contrast, it does not equalize the numbers of G's with A's, G's with T's, etc. because they are not paired with each other in the inverted segments.

The process is asymptotic because it is effectively irreversible. Every such inverted insertion brings the genome closer to perfect strand symmetry. After all, the equalization caused by a certain transposition/inversion can only be undone by reversing it exactly and immediately afterwards. Such an exact reversion, however, is extremely unlikely to occur in the random fashion, the transposition/inversions are assumed to happen.

The process is also self-stabilizing, because once a genome complies with Chargaff's second parity rule, the described mechanism maintains the compliance forever. In this case both strands of the inverted segment come from a genome that has - on average – already approximately equal numbers of complementary nucleotides. Thus as they are inverted and inserted somewhere else, they bring as many nucleotides into a strand as they took away from the complementary strand. Thus strand symmetry is a stable end state of genomes that are subjected to the process described by the transposition/inversion hypothesis.

Of course, one needs many such inversions/transpositions to achieve almost perfect strand symmetry all along a genome. However, that does not seem to be a major obstacle for the hypothesis. It seems safe to assume that genomes were subjected to many transposition /inversions in their evolutionary past. The very nature of transposable elements suggests the geometric growth of their numbers over time. Indeed, in some cases such as Alu, LINE and SINE sequences, millions of copies were found in human, mouse and other genomes, and they were found in coding and non-coding regions alike [13 - 16]. It is not known exactly how many of these transposons were inverted, but the tacit assumption in the field seems to be that they are on par with the non-inverted ones. Likewise, it is not known how many inversions any particular genome experienced. Yet, it seems reasonable to conclude that most sufficiently large inversions transferred some coding sequences from the Plus-strand to the Minus-strand and vice versa. Since most genomes contain coding regions on both strands, one may infer that they also experienced a large number of inversions in their past, even if they are no longer recognizable today.

The number of inversions/transpositions that a genome has experienced in its evolutionary past does not begin at the time when the species originated, but much earlier. When a species branched off the evolutionary tree its genome was formed from the mutation, concatenation and/or duplication of certain older genomes. Each of these precursor genomes had already experienced a large number of inversions/transpositions corresponding to their evolutionary age at the time of the branching. In other words, the number of inversions/transpositions that left their marks in a genome should be about the same, no matter how recently evolved a species is.

In short, the compliance with Chargaff's second parity rule may be interpreted as an inevitable, asymptotic product of - among other causes – numerous inversions and inverted transpositions that occurred in the course of evolution. The conversion of every initially non-compliant genome to a compliant one began presumably with relative small genomes like bacterial genomes, which gradually grew in size while at the same time the described mechanism - and the additional mechanisms described by earlier hypotheses - improved their degree of compliance with the rules. As the inevitable consequence of transposition/inversions the above mechanism changes all genomes indiscriminately. Therefore, the compliance of a genome with the rules seems to present no constraint and offers no selective advantage over less compliant ones.

As the described process is asymptotic in nature, no genome can ever become perfectly compliant by it. Nevertheless, as a genome experiences more and more transposition/inversions, their equalizing effect covers the entire length of the genome more and more completely. Thus, the residual areas where a genome still violates the rules must decrease steadily in size. In other words, consistent with the above results, the smaller a segment of a present day genome the more likely it may still violate the rules to some degree.

The Transposition/Inversion Hypothesis of the 'Majority Profile'

Expressing strand symmetry does by no means determine the triplet profile of a species. Mathematically speaking, the number of conceivable triplet profiles that fulfill Chargaff's second parity rule is astronomically large and could easily afford a differently shaped profile for every variant genome of every species. Thus, based on the huge variety of different species, one may expect a correspondingly large variety of the frequency distribution of their tri-nucleotides ("triplet profiles"). Yet, I found that there are essentially only 3 kinds of triplet profiles among the large number of genomes examined here. None of the classes included random profiles, all of them contained members from vastly different taxa and species. Therefore, the three classes of genomes do not reflect the phylogeny of their member organisms. Instead, they may reflect species-independent mechanisms of genome evolution.

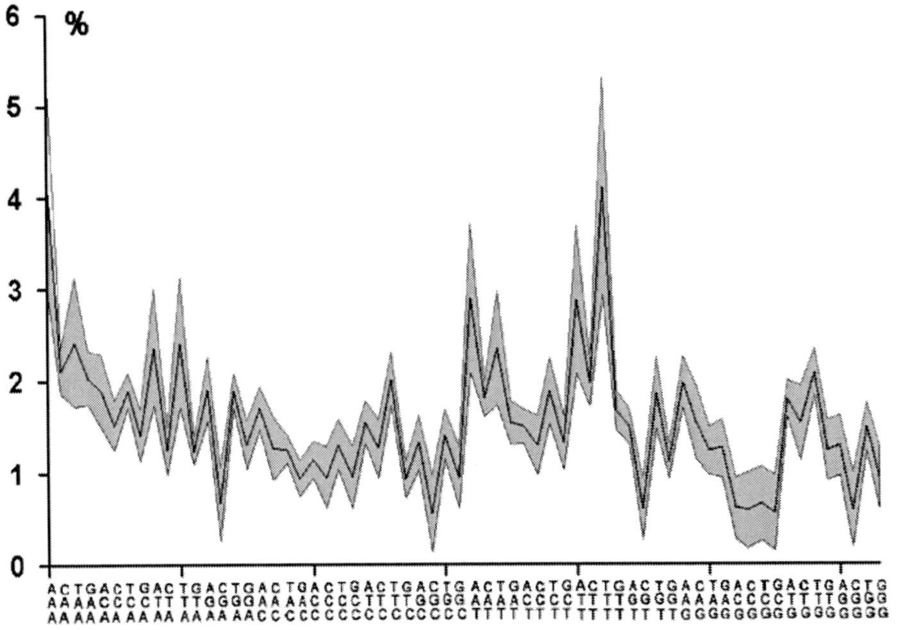

Figure 1. Majority Triplet profile (Abscissa: triplets to be read from bottom to top; Ordinate: fraction of triplets of entire genome). The shaded area covers the range of the standard deviation computed from the triplet profiles of 31 chromosomes of different organisms ranging from rickettsia to humans, including human, chimpanzee, mouse, zebrafish, maize, *streptococcus pneum*., *Arabidopsis*, *xenopus laevis*, yeast, *B. subtilis*, *anopheles*, and others. (Reprinted from Albrecht-Buehler G. (2007). Inversions and inverted transpositions as the basis for an almost universal 'format' of genome sequences. Genomics 90:297-305, with permission from Elsevier)

One class, called the 'majority class', contains the genomes of most tested organisms ranging from Rickettsiae to primates (Figure 1). Their frequency distribution $f_M(\Delta)$ which was called the 'majority triplet profile', or 'majority profile' for short, appears to be a surprisingly universal property of genomes regardless of taxa or species.

The following presents the arguments to propose that the same inversion/transposition mechanism(s), that created the strand symmetry, may also explain the existence of the majority profile. The arguments proceed in 3 steps (Figure 2).

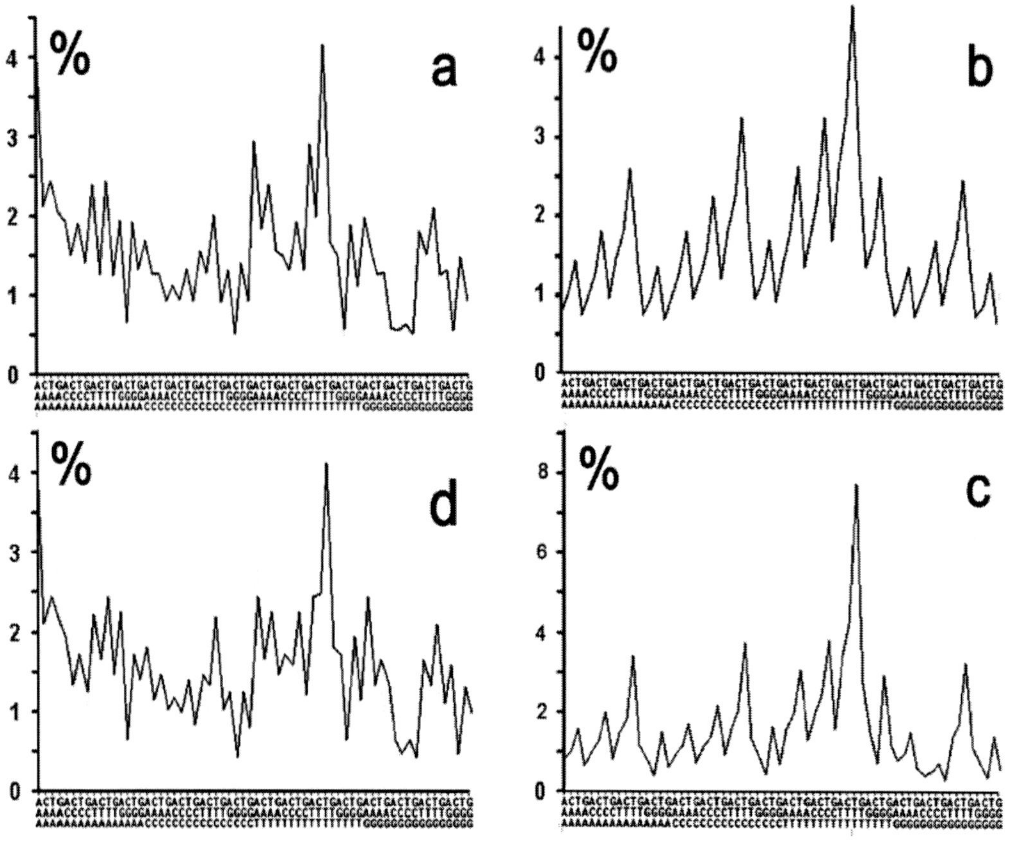

Figure 2. Rendering of triplet profiles similar to the majority profile by the procedure described in the text (Abscissa: triplets to be read from bottom to top; Ordinate: fraction of triplets of entire genome). a. The majority triplet profile. b. The stochastic expectation of the triplet profile resulting from an 'urn experiment with replacement' (equ.2) using the 'initial' frequencies p0(A) = 0.20, p0(T) = 0.36, p0(C) = 0.25, p0(G) = 0.19. c. Effect of replacing randomly 60% of all CG pairs with TT pairs on the 'genome' with the triplet profile of panel (b). d. Effect of 30,000 inversions/transpositions of 1 kb size on the triplet profile of the simulated genome of panel c. This triplet profile is very similar to the majority profile of panel a (correlation coefficient is 0.957). (Reprinted from Albrecht-Buehler G. (2007). Inversions and inverted transpositions as the basis for an almost universal 'format' of genome sequences. Genomics 90:297-305 , with permission from Elsevier)

Step1. The Stochastic Expectation

What kind of triplet profile can one expect, if genomes were completely random sequences (stochastic expectation)? In order to answer this question, we consider the sequence of nucleotides as the result of the random picking from an urn of well-mixed balls labeled as A, C, G, and T. Every selected ball determines the next nucleotide in the sequence, and is afterwards returned to the urn. This method guarantees that the sequence of the nucleotides is truly random, and that the probabilities of picking any of the four does not change as the sequence elongates.

Assume that the probabilities of the balls are p_0 (A), p_0 (T), p_0 (C), and p_0 (G). Whatever they are individually, it is guaranteed to pick one of them. In other words, the probability to pick one of them is always 100%:

$$p_0(A) + p_0(T) + p_0(C) + p_0(G) = 100\%. \quad (1)$$

In the case of the actual genomes of the majority class one can easily calculate from the majority profile their base compositions as $p_M(A) = 30.6\%$ $p_M(T) = 30.6\%$ $p_M(C) = 19.4\%$ $p_M(G) = 19.4\%$. It corresponds to an AT content of 61.2% and a CG content of 38.8 %. Obviously, these genomes obey Chargaff's second parity rule for mono-nucleotides, i.e. $p_M(A) = p_M(T)$ and $p_M(C) = p_M(G)$ on each single strand.

If the sequence was truly random then the probability to pick (say) an 'A' would have to be completely independent of the particular nucleotide that was picked right beforehand. Thus the probability $f_0(\Delta)$ of drawing the triplet $\Delta = \{n_1,n_2,n_3\}$ at random from the urn is given by the (the probability to pick the first) x (the probability to pick the second)x(the probability to pick the third):

$$f_0(\Delta) = p_0(n_1) \bullet p_0(n_2) \bullet p_0(n_3). \quad (2)$$

In general, this triplet profile will be very different from the majority profile. For example, if the probability of every nucleotide was the same, i.e. $p_0(A) = p_0(T) = p_0(C) = p_0(G) = 25\%$, then the triplet profile would be the same value $f_0(\Delta) = 1.56\%$ for all 64 possible triplets. Consequently, if depicted in the same way as in Figure 1 it would be a straight line parallel to the x-axis at a value of 1.56%. Obviously, the majority profile is very different from that constant function.

Assume next that the probabilities of all four bases are different, yet still obey equ. (1). The resulting sequence will still be random, because picking the Nth nucleotide that follows a sequence of N-1 nucleotides remains a stochastic event that does not determine which particular nucleotide it will be. Mathematically speaking, the resulting sequence would be called a Markov chain of zero order. For example, let the probabilities (somewhat arbitrarily) be

$$p_0(A) = 20\% \ p_0(T) = 36\% \ p_0(C) = 25\% \ p_0(G) = 19\%. \quad (3)$$

Note that only 3 parameters are assumed in equ. (3) because the 4th is determined by equ. (1). The effect of the choice of the particular parameters on the results is discussed elsewhere [3].

A specially designed computer program created 8 Mb large, random genome sequences with the particular base composition of Equation 3 and confirmed by direct counting that their triplet profile (Figure 2b) matched exactly the prediction of Equation2.

Such artificial genome sequences represent the stochastic expectation of genomes randomly arising from a large pool of nucleotides as e.g. postulated for example by the so-called "RNA-world" [19-23]. Obviously, there are astronomically large numbers of such genomes that have different sequences yet, nevertheless, have the same triplet profile of Equation 3. However, the genome sequences so generated do not fulfill Chargaff's second parity rule, nor do they belong to the majority class (see Figure2b).

Step2. Sequence Modification

CG-to-TT Conversion

According to our previous results, applying a large number of inversion/transpositions to the stochastic-expectation genomes must generate sequences that fulfill Chargaff's second parity rule [1]. However, this process did not immediately yield triplet profiles that resembled the majority profile. Therefore, it was necessary to modify the stochastic-expectation genomes beforehand in a certain way. I found that the simplest modification yielding a majority profile after inversions/transpositions was to change 60% of CG di-nucleotides into TT di-nucleotides. This modification was expressed by a 'conversion rate' of 0.6.

Even though a conversion rate of 60 % may sound like a large modification, the number of sequence alterations was actually rather small. Using the base composition of Equation 3, the stochastic-expectation genomes contained approximately 5% CG pairs on either strand. Therefore, the required sequence modification involved only $0.6 \bullet 5 = 3\%$ modification of their total di-nucleotides. As a result, the triplet profiles that resulted from this modification (Figure2c) were qualitatively not very different from the initial profiles, although certain peaks were increased (Note the change of scale in Figure2c). At this stage, the modified sequences still violated Chargaff's second parity rule.

I found no precedence for a CG-to-TT conversion in the literature. However, it may be interpreted as the result of the following 2-step process, for which there is precedence in contemporary genomes. The first step may involve a CG-to-TA conversion for which there is ample precedent in the literature, because methylated C is frequently converted into T [24]. Given a CG pair on (say) the Plus-strand, base pairing requires that there is a CG pair opposite to it on the Minus-strand. (Note that both strands are read in the 5' to 3' direction). Converting both (presumably methylated) C's into T's would generate a pair of mismatched TG di-nucleotides. However, during the next round of replication they could be corrected into TA pairs [25,26].

The second step could be viewed as an A-to-T substitution that would turn some of the newly generated TA pairs on (say) the Plus-strand into TT pairs. Correspondingly, their reverse complements on the Minus-strand would turn into AA pairs. Such A to T substitutions have been observed in a number of cases [27]. Still, the above is merely a plausibility argument, since cytosine methylation is not a ubiquitous process. Therefore, it is quite possible, and perhaps even likely, that the postulated CG-to-TT conversion occurred by some other, yet to be discovered mechanism.

Alternative: Markov Chains of Higher Order

The absence of any known mechanism that could achieve the CG-to-TT conversion together with the required high percentage of conversions may make it desirable to find an alternative mechanism that would accomplish the same sequence modification.

One such possibility would be to consider a Markov chain of the first (or even higher) order. In this case the next (N^{th}) nucleotide in a forming chain of (N-1) bases would no longer be determined only by its availability in the 'soup' (equ. (2), but also by the specific nucleotide that occupies the last, namely the $(N-1)^{st}$ position in the chain.

In order to apply this concept one has to define additional parameters, called transition probabilities $p_X(Y)$. For example, one would need to define the probability $p_C(A)$ which indicates how likely it is that the Nth nucleotide is an 'A' if the previous one was a 'C'. Similarly, $p_G(T)$ would indicate how likely it is that the Nth nucleotide is a 'T' if the previous was a 'G', and so forth.

In effect the successor of a base X in the chain is now no longer determined by a single base frequency, but by four different ones, namely one distribution $p_X(Y)$ for each fixed base X. Each is independently normalized to 100%.

$$p_X(A) + p_X(T) + p_X(C) + p_X(G) = 100\%. \quad (X = A,C,T,G) \qquad (1')$$

Thus each base enforces its own specific base frequency distribution which determines the probability for its successor in the chain.

On the face of it, there seem to be 16 transition probabilities $p_X(Y)$ with (X = A,C,T,G and Y = A,C,T,G) that need to be defined. Actually only 12 are needed because equ (1') binds 4 of the variables. Once all transition probabilities are defined, equ. (2) becomes ($\Delta = \{n_1,n_2,n_3\}$)

$$f_1(\Delta) = p_0(n_1) \bullet p_{n1}(n_2) \bullet p_{n2}(n_3). \qquad (2')$$

Note that the first factor is not a transition probability, but an overall base frequency. This base frequency is no longer a freely selectable parameter as in the case of a zero order Markov chain, but it is determined by all transition probabilities, and has to be calculated from them in the following way. Defining

$$\pi(Y) = p_A(Y) + p_C(Y) + p_T(Y) + p_G(Y). \qquad (Y= A,C,T,G) \qquad (4a)$$

the normalized overall base frequency p(Y) can be computed as

$$p_0(Y) = \pi(Y) / [\pi(A)+\pi(C)+\pi(T)+\pi(G)] \qquad (Y= A,C,T,G) \qquad (4b)$$

The advantage of using a first order Markov chain approach is the possibility to skip the stage of the initial poly-nucleotide as depicted in Figure 2b and assume that they formed out of a 'primordial soup' right away with triplet profiles like the one depicted in Figure 2c.

There is a disadvantage, however, because one needs to define a much larger number of parameters than in the case of the CG-to-TT conversion.. Worse, one needs to *explain biochemically or otherwise all of them.* For example, one would have to explain why in the 'primordial soup' the probability $p_G(T)$ that indicates how likely it is that a 'T' follows a 'G' is different from the probability $p_G(G)$ that indicates how likely it is that a 'G' follows a 'G'. One could think of explanations that invoke steric hindrance for the addition of certain nucleotides after certain others. Alternatively, one may think of mechanisms of chain elongation which add entire pairs of nucleotides that are preformed in solution, and thus make the appearance of sequential pairs of certain nucleotides in the sequence more likely. Of course, these and other interpretations of the transition probabilities have to be logical, be based on experiments, and have to be consistent with other situations where nucleotides form chains. For example, if one wants to hold something universal such as steric hindrance responsible for the transition probabilities derived from the human genome, one would also have to prove that the same transition probabilities apply to every other genome. After all, the steric hindrance could not be specific for humans.

It is not difficult to come up with suitable parameters for such a model. As mentioned above, there are 12 free parameters available to fit the 63 values of triplet profile of Figure 2c. (Actually, the function has 64 values, but the normalization requirement binds one of them). Anybody familiar with fitting procedures will concede that it is quite easy to do the fitting with such a relatively large number of fitting parameters, especially, since the desired end result, namely the majority profile, has some variability as seen in Figure 1. However, as mentioned above, the difficulty may arise from the necessity to explain their values. Thus our attempt to avoid having to explain the one unknown mechanism of the CG-to-TT conversions leaves us with the requirement to explain biologically 12 other parameters that ultimately describe the unknown polymerization conditions for nucleotides in the 'primordial soup'.

Obviously, at the present time the knowledge about the origin of poly-nucleotide is not available to make a decision which of these and possibly other models should be used to explain the triplet profile of Figure 2c. One should also keep in mind that the credibility of an explanation is the better the fewer parameters and assumptions it requires. Regardless, however, what the final explanation will be, we still need the actions of numerous inversions/transpositions to turn it into the majority profile as described in the following section.

Step 3. The Intuitively Obvious Effect of Numerous Inversions and Inverted Transpositions

The final step subjected the modified genomes to N_{inv} = 30,000 inversions/transpositions (transposon size = 1 kb). That step was sufficient to transform their triplet profiles into the profile shown in Figure2d. This profile appeared very similar to the majority profile (correlation coefficient 0.96), and since it was the result of inversions/transpositions it naturally fulfilled Chargaff's second parity rule quite accurately.

The action of 30,000 inversions/transpositions may seem rather obscure because it happens inside the impenetrable 'bowels' of a computer and, thereby, seems to offer little explanatory power. However, in this case the result is quite easy to understand even without the help of a computer.

As shown in the earlier article [1], the effect of a large number of inversions/transpositions is to equalize asymptotically the frequency of every triplet with the frequency of its reverse complement in an exponential fashion. For example, before the application of inversions/transpositions, the normalized frequency of the GAA triplet was $f_0(GAA) = 0.6\%$ and that of its reverse complement was $f_0(TTC) = 3.8\%$). After a sufficiently large number of inversions/transpositions, both frequencies will became the same, namely $f_\infty(GAA) = f_\infty(TTC) = (0.6 + 3.8)/2 = 2.2\ \%$. In other words, *the effect of the inversions/transpositions is nothing more than to generate the arithmetic means between the frequencies of triplets and their reverse complements.*

Denoting the reverse complement of a triplet Δ by the symbol ▼ this result can be formulated as

$$f_\infty(\Delta) = f_\infty(\blacktriangledown) = [f_0(\Delta) + f_0(\blacktriangledown)]/2 \qquad (5)$$

The inversions/transpositions do not only equalize the triplet frequencies, but the frequencies of mono-nucleotides, di-nucleotides, etc. as well [1]. According to equs. 3 and 4, the equalization of mono-nucleotides would yield $p_\infty(A) = p_\infty(T) = [p_0(A) + p_0(T)]/2 = 28\%$, and $p_\infty(C) = p_\infty(G) = [p_0(C) + p_0(G)]/2 = 22\%$. Note that these frequencies differ from the frequencies $p_M(X)$ of the majority profile. However, the difference is adjusted by the CG->TT conversions.

In view of this simple algorithm the accurate reproducibility of the effect of numerous inversions/transpositions is self-evident. Therefore, it seemed unnecessary to document the reproducibility of the results by statistical significance calculations. Nevertheless, I repeated the generation of the majority profile as described more than 20 times with identical results.

B. Genome Signatures

Based on the above background of the universal strand symmetry, the almost universal majority profile, and the role of inversions/transpositions in the creation of these universals, we are now in the position to examine the deviations of a genome from them. To be sure, there should be deviations regardless of how universal these properties are, if only because the universals are the result of an asymptotic process which should leave residual areas of not yet perfect expression.

As will be shown below, there are, indeed, many genome segments that deviate considerably from these universals, and especially from the majority profile. Most surprising, however, is the demonstration that countless major deviations were created *de novo* over and over again during the course of evolution. Obviously, these deviations must be the result of an unknown mechanism that acts as some kind of antagonist of the actions of inversions/transpositions which continue to erase such deviations. Still, even the specific expression of this antagonist is shaped by inversions/transpositions to a large degree, as will be shown below, as well. The latter finding, perhaps more than any other presented here, emphasizes how powerful and important inversions/transpositions seem to be for the format and sequence architecture of genomes.

Signatures: Local Deviations from Genome Universals

If evaluated in their entirety, most genomes expressed the strand symmetry and the similarity of their triplet profile with the majority profile to an amazingly high degree of perfection. However, not every part of every genome expressed it to the same degree. Using a small probe segment to scan continuously along a genome and testing its local deviation from these global genome properties yielded quite characteristic patterns. They will be called 'signatures'

The preceding sections argued that numerous inversions /transpositions in the evolutionary past of every genome have improved asymptotically its strand symmetry and its similarity to the majority profile. Therefore, one may be tempted to make some simple predictions about both kinds of genome signatures. For example, one might expect that they are the more leveled out the older the species, as its genome had more time to approach its end state. Alternatively, recognizing that the genomes of all contemporary species are actually equally old, one may expect that all signatures show only minute deviations from their end states. After all, a billion years or more of evolution should suffice to reach this end state quantitatively. Yet, none of these expectations were substantiated by the analysis reported here.

Profile Signatures

The scanning procedure used here consisted of selecting a probe segment of 100 kb and advancing it along the entire genome in steps of 10 kb. At every step the correlation coefficient between the triplet profile of the probe and the majority profile was measured. The procedure yielded a pattern that will be called the *'profile signature'* of the genome. The present chapter suggests four properties of profile signatures that stand in stark contrast to the above mentioned expectations.

1 The profile signatures of most species showed dramatic patterns of 'bursts' and 'flat regions' throughout the length of each chromosome (e.g. Figure 3b).
2 Descendent species seemed to inherit large parts of their profile signatures from their ancestors. Therefore, significant changes of profile signatures may be indicators of significant evolutionary change.
3 Profile signatures were not simply leveled out during evolution as one might expect from the mentioned recent articles [1, 3], but that frequently new ones were created. The present chapter argues that the times of new bursts may mark branching points of the phylogenetic tree.
4 The new bursts seem to be the result of a universal mechanism that changes locally the AT contents. This conclusion is suggested by a universal 'sigmoid curve' that describes the correlation between the amplitude of profile signatures and the local AT contents of the genomes.

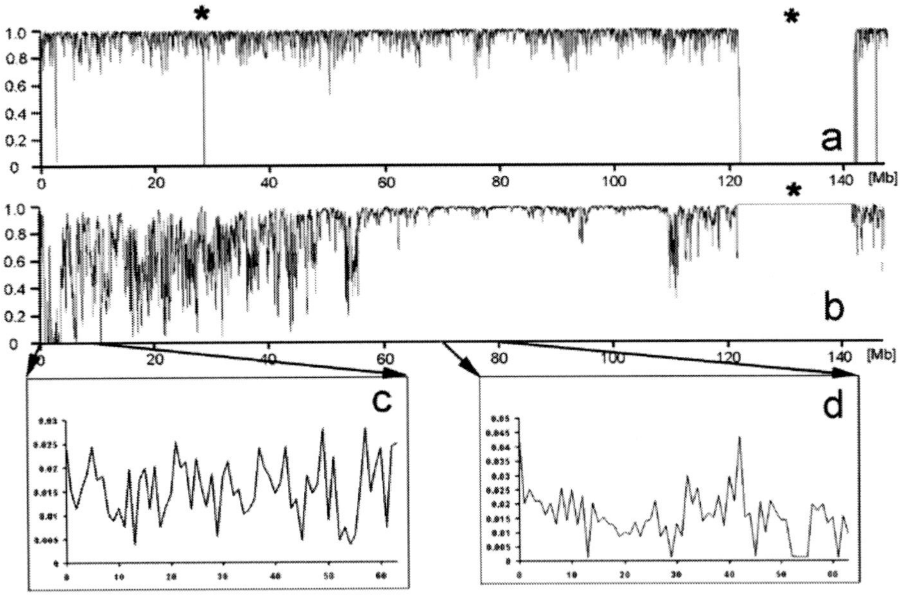

Figure 3. Illustration of genome signatures(= local deviation from a particular genome measure determined in a probe segment of 100 kb advancing in steps of 10 kb along the genome sequence starting at the 3' end) in the case of the first 145 Mb of human chr 1. An asterix indicates repetitive DNA that was masked by N's or n's in the genome sequence. (Abscissa: position along genome [Mb]; ordinate: fraction of deviation every 10 kb). a. Strand symmetry signature (= local deviation from the strand symmetry, described as Chargaff's second parity rule). b. Profile signature (= local deviation from the majority profile). c. Local triplet profile in the area of a 'burst' (left 2 arrows), which is very different from the majority profile. d. Local triplet profile in a 'flat' area (right 2 arrows), which is identical to the majority profile.

Strand Symmetry Signatures

In addition, I measured the degree of strand symmetry (i.e. the correlation coefficient between the triplet profiles of the Watson- and the Minus-strand) using a similar procedure. The resulting pattern of this measure will be called the *'strand symmetry signature'* of that genome. The strand symmetry signatures of the genomes tested were more in compliance with the above mentioned expectations. They showed only small, ‘noisy’ amplitudes without any characteristic patterns that distinguished the genome of one species from another. Therefore, the present chapter will not elaborate on them (Figure3a)

In contrast, the profile signature changed quite dramatically along the length of human chromosome 1 (Figure3b). This finding, too, was quite typical for most other chromosomes of this and other members of the majority group. There were many 'bursts' in the form of an extended series of large amplitudes of deviation from the majority profile (Figure3c) alternating with long stretches where the local triplet profile was almost identical with it ('flat portions'; Figure3d).

Conservation of Profile Signatures

The pattern of alternating bursts and flat portions appeared to be highly conserved for millions of years. An example is shown in Figure4 for chromosomes 1 of humans, chimpanzees and rhesus monkey.

It is apparent that the signatures of these primates were very similar to each other in spite of the large evolutionary time differences between them. A closer inspection showed that human chromosome 1 compared to chimpanzee and rhesus had additional stretches inserted in various places. Therefore, the similarities are sometimes more obvious if one stretches the graphic displays of the profile signatures to the same size, as was done in Figure 5.

In other cases local inversions have occurred. For example, the burst in rhesus chromosome 1 around position 160 Mb appears inverted compared to the corresponding bursts at location 180 Mb in chimpanzee chromosome 1 and 200 Mb in human chromosome 1 (Figure 5).

Since each curve shown in Figure4 condenses more than 20,000 data points into the width of a page, one may suspect that this similarity is confined to low resolution depictions. However, the similarity extended to considerably higher levels of resolution as shown in Figure5, which amplifies the resolution of two corresponding stretches of chromosomes 1 of humans and chimpanzees twenty-fold.

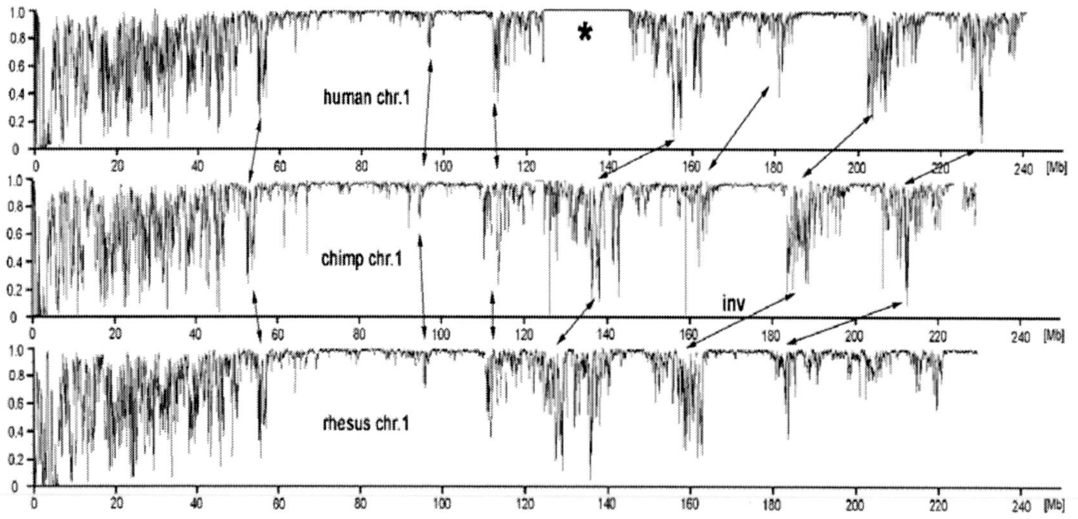

Figure 4 .Example of the conservation of profile signatures. (Axes as in Figure3b). Arrows between the profile signatures of chromosome 1 of three primates (human, chimpanzee and rhesus monkey) point to patterns of bursts with a high degree of similarity. Arrows point to corresponding groups of bursts. Note: Each of these signatures containing approximately 20,000 data points include coding and non-coding regions alike.

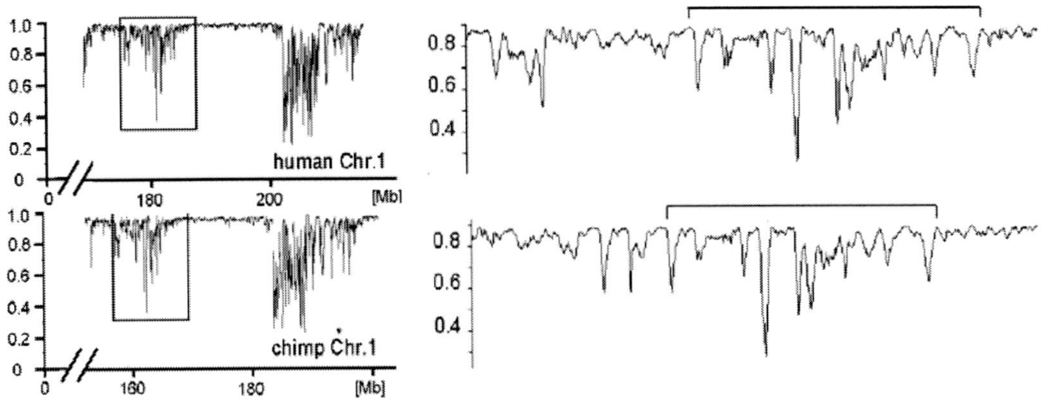

Figure 5. Conservation of profile signatures at a higher level of resolution. The framed groups of corresponding bursts in chromosomes 1 of human and chimpanzee are shown enlarged about 20 times in the right hand panels. The brackets indicate almost identical stretches of the profile signatures.

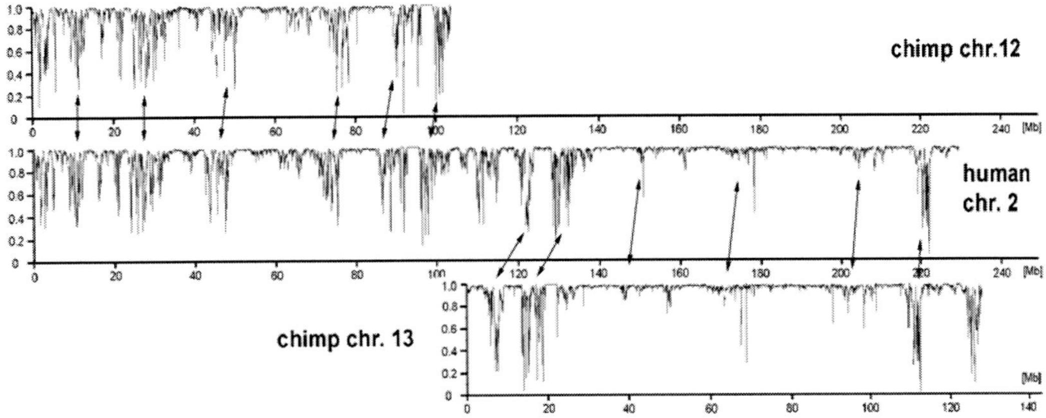

Figure 6. Profile signatures as markers of chromosome concatenation. (Axes as in Figure3b). The arrows between profile signatures of human chromosome 2 and chimpanzee chromosome 12 and 13 suggest that the former is very similar to the concatenation of the 2 latter. This fact, well-known through the compilation of sequence homologies, can be much more rapidly established by inspection of profile signatures that also include non-coding regions. Arrows connect corresponding bursts.

The result may not seem too surprising, considering the well-known high degree of sequence homology between genes of humans and chimpanzees. However, it should be noted, that profile signatures include also the entirety of the non-coding regions, in which sequence homologies are not as well established as in the coding regions.

Profile Signatures as Markers of Chromosome Rearrangements and Synteny

The high degree of conservation should qualify profile signatures as markers of chromosome rearrangements. In order to test this, I selected several well-known cases of chromosome concatenation and inversion and examined how their profile signatures would reflect this fact.

Figure 6 shows the case of the concatenation of chromosomes 12 and 13 of chimpanzee, which formed the human chromosome 2. The concatenation, known from synteny charts, is quite obvious in the profile signatures of the chromosomes. Likewise, a chromosome inversion such as a large portion of mouse chromosome 5 and rat chromosome 14 are quite easily recognized by their profile signatures (Figure 7).

In general, the profile signatures of rat and mouse revealed a high degree of conservation (e.g. Figure 7). In contrast, I found no general similarities between the profile signatures of these rodents and the mentioned primates except in the well-known area of synteny between human chr.1 on one hand and the inverted chr.4 and chr.5 of mouse and rat, respectively (Figure 8).

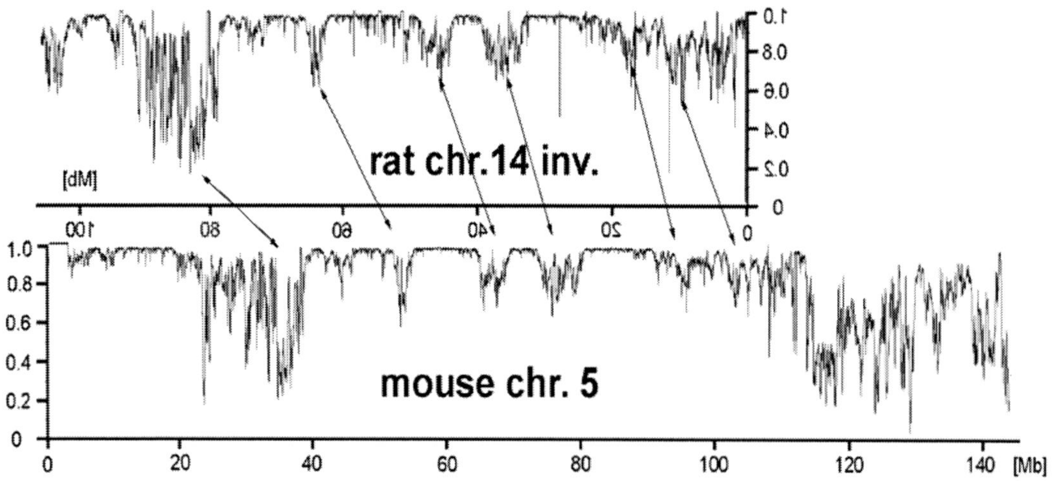

Figure 7. Profile signatures as markers of chromosome inversion. (Axes as in Figure 3b). The arrows between the profile signatures of mouse chromosome 5 and rat chromosome 14 show that a part of the former is very similar to the

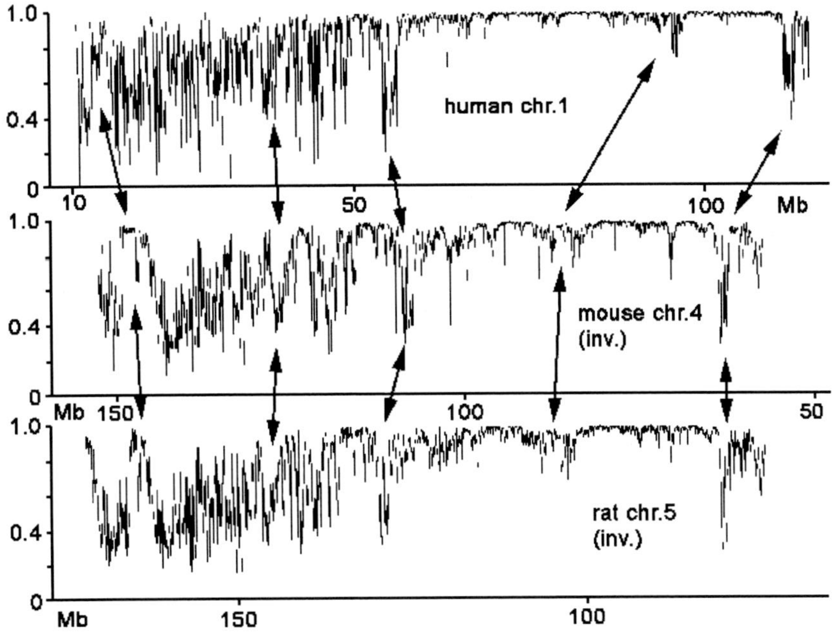

Note: Chromosomes with the same number do not necessarily have related profile signatures. For example, a case like the concatenation of 2 small chromosomes into a large one (e.g. Figure 6) changes the order of chromosome sizes and thus their numbers.

Figure 8. Example of synteny expressed in profile signatures (Axes as in Figure 3b). Corresponding bursts of the profile signatures of human chr.1 (10 -129 Mb), the inverted mouse chr. 4 (154 - 58 Mb), and the inverted rat chr. 5 (171 - 75 Mb) are connected by arrows.

To date, the demonstration of such syntenic relationships between chromosomes is based on the rather time consuming compilation of sequence homologies. In contrast, it takes only minutes to establish the profile signature of a several hundred Mb large chromosome if its sequence is known. Therefore, profile signatures my offer a quick way to gain overview over potential candidates for synteny. Subsequently, the results can be confirmed by more targeted and detailed tests of sequence homologies.

Total Fraction of Bursts (TFB)

As mentioned earlier, the depictions of profile signatures is highly condensed in the above figures. Due to the relatively thick lines of the graphs the total fraction of a genome covered by bursts may appear much higher than it actually is. For a more accurate measurement of that fraction, I added up the cumulative size of the 10 kb large test steps where the profile signature of the entire genome of an organism dropped below 0.8 and divided it by the size of the total genome. In order to examine examples of members of different families, classes and phyla, the results were derived from the complete genome sequences of opossum, human, chimpanzee, rhesus, rat, mouse, platypus, arabidopsis, maize, sea squirt, zebrafish, fugu, C.elegans, and C. briggsae.

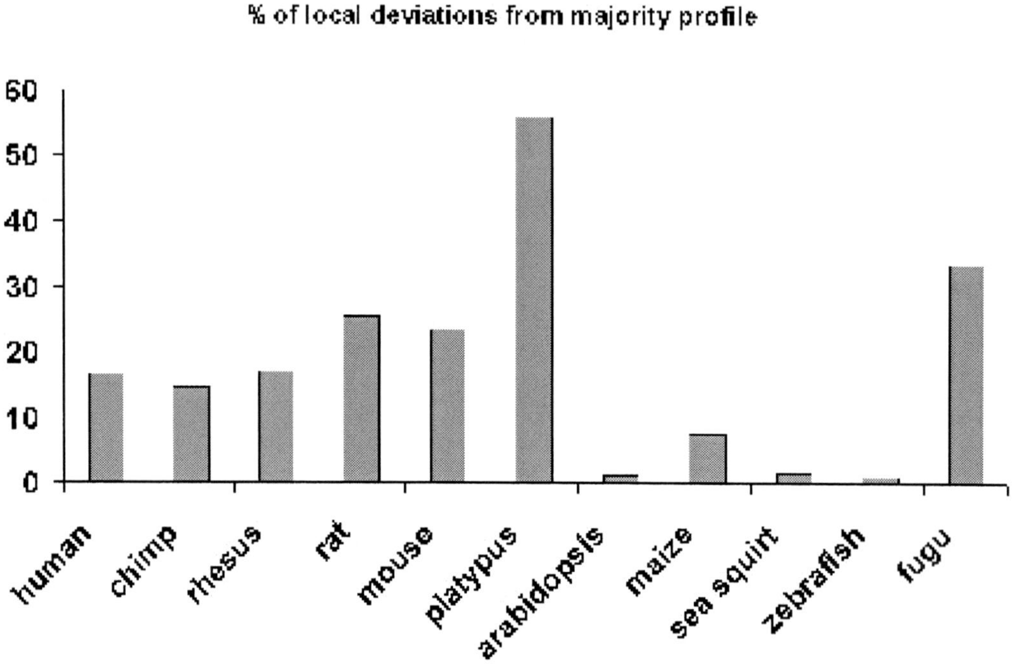

Figure 9. Total fraction of bursts (TFB). The figure shows the fraction of total genome length where the profile signature of the various organisms, indicated at the abscissa, dropped below 0.8. The data comprise the entire genomes of these organisms and not merely selected chromosomes.

Figure 9 shows that the TFB value of profile signatures was highly variable, and obviously unrelated to the evolutionary age of the corresponding organism. The only possible relationship between the data appeared to be a similarity between members of the same order.

Profile Signatures as the Result of Local AT Changes (Universal Sigmoid Curve)

I also measured the local AT content of each probe as it scanned along the tested genomes. (One might call the result an "AT signature", but it seems unnecessary to introduce yet another new term.) A profile vs. AT-content correlation plot between amplitudes of bursts in the profile signature and the local AT contents yielded a surprisingly well-defined 'sigmoid curve'. Figure 10 shows the typical example of a burst of 10 Mb size at the beginning of human chromosome 7 (Figure10a).

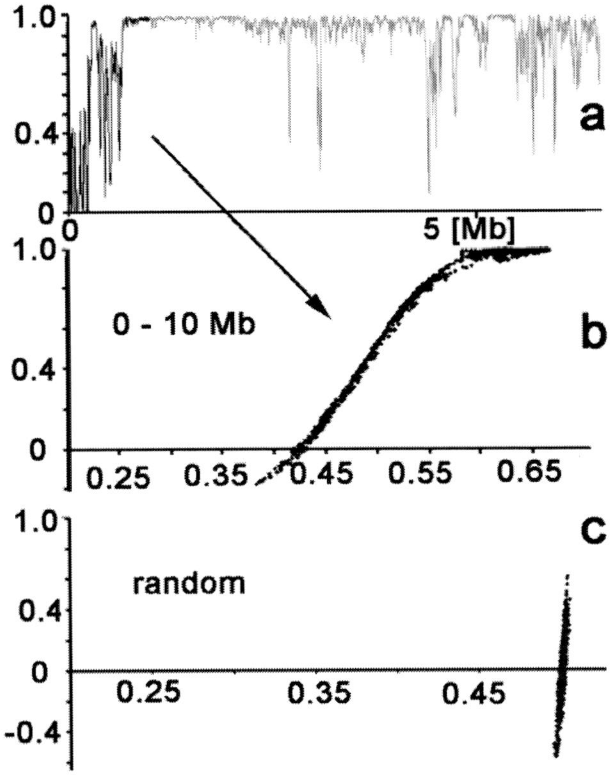

Figure 10. The sigmoid curve. (Abscissa: AT-content measured in probe segments of 100 kb that advanced in steps of 10 kb along a section of a genome; ordinate: deviation of the same probe segment from the majority profile.) a. Profile signature of part of human chromosome 7 with a highlighted burst of 10 Mb size at its beginning. b. Correlation plot between the amplitudes of the highlighted burst and the local AT contents yielding the typical shape of the sigmoid curve. c. For comparison the same plot of a randomly constructed genome. As the AT content of each 100 kb large segment of the random file is essentially constant, the plot yields an approximately vertical line with its data points located symmetrical around the abscissa.

The sigmoid curve indicated that the bursts were not as random as they seemed, but appeared linked to a single variable with little stochastic 'noise' (Figure 10b). Furthermore, the bursts did not create random nucleotide sequences, either. As shown in Figure 10c, a similar correlation plot of a computer-generated random sequence was very different. Since its local AT content was essentially constant, it yielded a straight, almost vertical line of data points distributed symmetrical about the AT-axis.

The sigmoid curve turned out to be identical for different members of the same class. Figure11a shows superimposed sigmoid curves of human chromosome 1, chimpanzee chromosome 1, mouse chromosome 2, rat chromosome 1, opossum chromosome 1 (0 - 320 Mb), and platypus (0 - 253 Mb). In spite of the large number of approximately 140,000 data points the sigmoid curves superimpose rather accurately along the same graph.

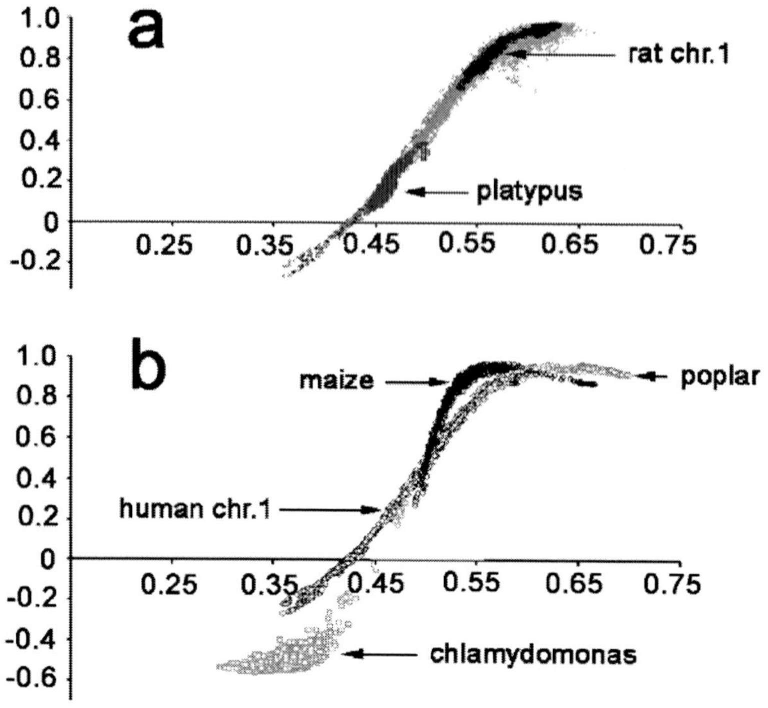

Figure 11. Degree of conservation and universality of the sigmoid curve. (Abscissa: AT-content measured in probe segments of 100 kb every 10 kb along a genome; Ordinate: Similarity between triplet profile and the majority profile of the same segments expressed as correlation coefficient between them). a. Conservation of the sigmoid curve among mammals. Superimposition of the sigmoid curves of human chromosome 1, chimpanzee chromosome 1, mouse chromosome 1, rat chromosome 2, opossum chromosome 1 (0 - 320 Mb), and platypus (0 - 253 Mb). In spite of the large number of 140,000 data points, the resulting superimposition remains quite well defined. Please note that platypus does not belong to the majority group and, yet, its data points fall on the same sigmoid curve as other mammals. b. Differences between the sigmoid curves from genomes that belong to different phyla and kingdoms. Although all yield sigmoid curves, they differ in the slope and location of their inflection point. Please note that chlamydomonas does not belong to the majority group and, yet, its data points seem to form a part of a sigmoid curve.

In contrast, the sigmoid curves of members of other phyla and kingdoms could differ markedly in the slope and location of their inflection points as shown in the examples of Figure 11b.

Relatively flat profile signatures of genomes or parts of genomes can cover only a small part of the sigmoid curve. Yet, these parts coincided quite well with the curve. Even flat profile signatures of genomes that did not belong to the majority group such as platypus (highlighted in Figure11a) and chlamydomonas (marked in Figure11b) coincided with the curve quite accurately. The latter genomes have an overall high GC content; yet, their data points seemed to fall as much on parts of the same sigmoid curve as members of the majority group. It suggests that the sigmoid curve expresses more than merely a special property of genomes of the majority group.

Another example of this larger than expected universality of the sigmoid curve resulted from the examination of 50 mitochondrial genomes. They belonged to different organisms from a wide range of kingdoms and species. Their generally small genome size precluded a determination of their profile signatures by the described scanning protocol. Therefore, I plotted their overall deviation from the majority profile versus their overall AT content. Although many of these genomes did not even comply with Chargaff's second parity rule, their data points seemed to fall on a sigmoid curve that was shifted by 0.06 towards higher AT contents (Figure 12), nevertheless. Plotting the overall deviation of 50 different mitochondrial genomes from the majority profile versus their overall AT content suggested also a sigmoid relationship (marked by filled diamonds). Although the selected genomes belonged to different organisms from a wide range of kingdoms and species, and many of them even violated Chargaff's second parity rule, their data points seemed to fall on a sigmoid curve that was shifted by 0.06 towards higher AT contents. (For a better comparison the sigmoid curves of Figure 8b were added with reduced contrast.)

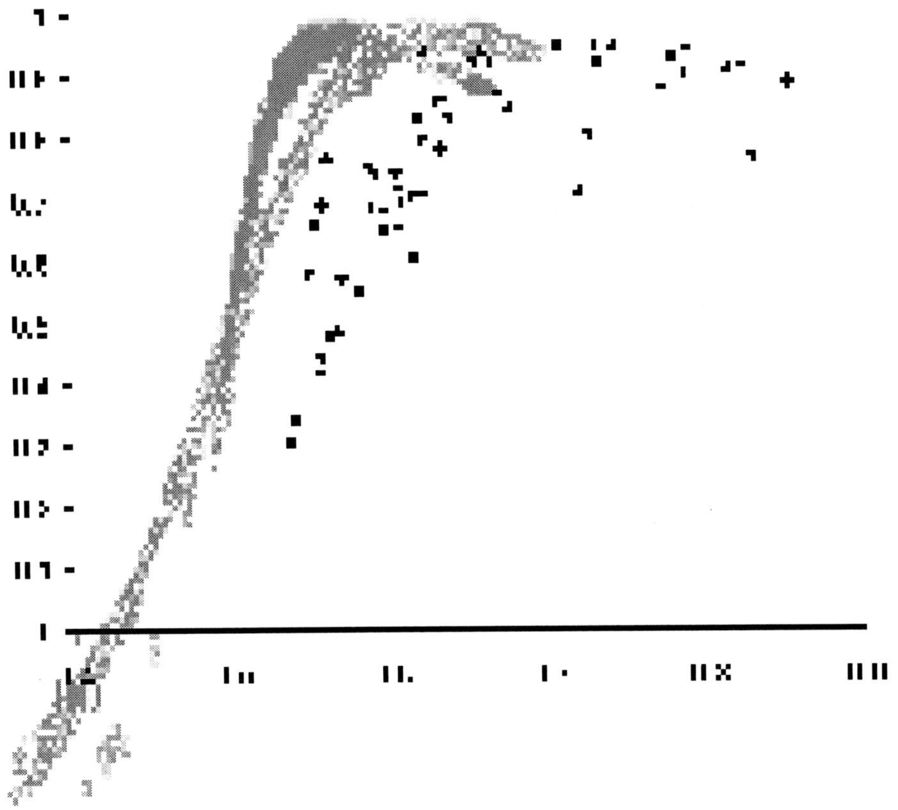

Figure 12. Universality of the sigmoid curve. (Abscissa: AT-content measured in probe segments of 100 kb every 10 kb along a genome; Ordinate: Similarity between triplet profile and the majority profile of the same segments expressed as correlation coefficient between them).

In summary, bursts have the following properties.

1 They are local phenomena that involve stretches of 10 – 20 Mb length.
2 They result in a local decrease of the AT content compared to the AT content of the genomes of the majority group.
3 The reduction of the local AT content follows the universal sigmoid curve.

Simulation of the Sigmoid Curve

The narrowness of the sigmoid curve suggests that the bursts of profile amplitudes depend essentially on only one parameter, namely the local AT content. This is surprising, because the correlation with the majority profile, be it local or global, should depend on many more variables. After all, the triplet profile of a genome is not determined by its four base frequencies, let alone by its AT content.

Therefore, the narrowness of the sigmoid curve points to the existence of a special mechanism which alters triplet frequencies of genomes. It appears to act by changing AT contents locally in a particular way throughout the genome in coding or non-coding domains alike. Furthermore, the universality of the resulting sigmoid curve suggests that this mechanism is universal, as well.

At present, this mechanism is unknown. It is not even known at this time, whether the bursts resulted from new insertion into genomes or whether they were mutations of pre-existing parts. To be sure, they were certainly not caused by inversions/transpositions. After all, due to base pairing the local AT content has to be the same on both strands of every segment. Therefore, moving and inverting a segment of DNA from some other part of the genome cannot change its local AT content and cause a local deviation from the majority profile if the whole genome is AT rich.

However, computer simulations suggested that the effect of this unknown mechanism may not amount to much more than the conversion (replacement) of A's and T's with the same number of C's and G's in random places of the sequence. This is illustrated in Figure13.

For the simulation I selected a stretch of genomic sequence with a flat signature such as chimpanzee chromosome 14 (position 32 Mb - 40 Mb) that showed almost no deviations from the majority profile (Figure13a). Subsequently, the sequence was subjected to p = 5, 15, and 25% random substitutions of A's and T's with G's and C's. For example, a random substitution of p = 5% meant that 5% of A's were replaced with C's and 5% of T's with G's in randomly chosen locations of the sequence. Figure13b shows that the overall similarity of the resulting sequence with the majority profile decreased along the upper portion of the sigmoid curve with increasing amounts of random substitutions.

How then does the sigmoid curve arise? Consider a long stretch of genome sequence whose profile is similar to the majority profile. Therefore its profile signature is a flat line at a value of approximately 1.0 that runs parallel to the x-axis similar to Figure 13a. Now let a shorter fraction of it become subject to the described reduction of the AT-content with a matching increase in GC-content. According to the above simulation, the profile signature of this fraction becomes a burst, and its profile vs.

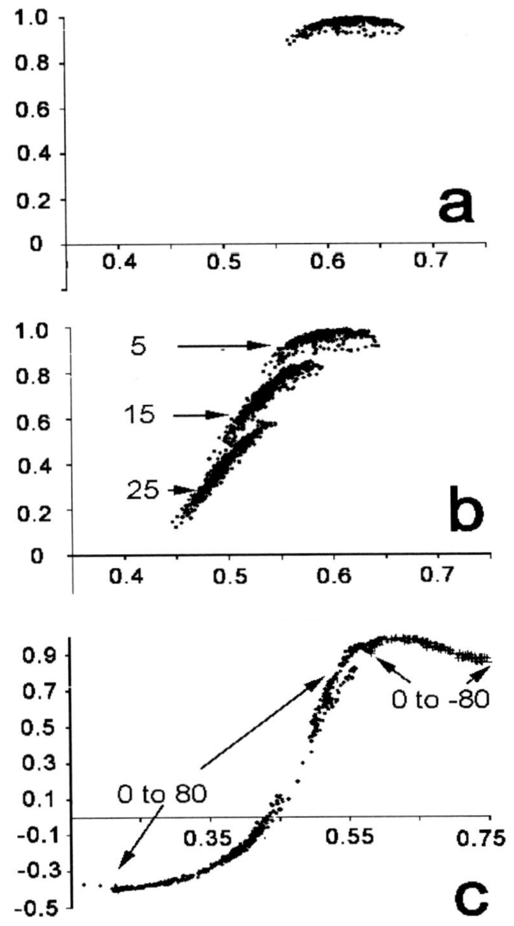

Figure 13. Simulation of the sigmoid curve by a substitution of A's and T's with C's and G's at equal numbers in random places of the genome sequence. (Abscissa: AT-content measured in probe segments of 100 kb every 10 kb along a genome; Ordinate: Similarity between triplet profile and the majority profile of the same segments expressed as correlation coefficient between them). a. Sigmoid curve of a flat stretch of chimpanzee chromosome 14 (position 32 Mb - 40 Mb) .b. Sigmoid curve of the same stretch after subjecting it to random substitutions of A's and T's with C's and G's in the amounts of p = 5, 15, and 25%. For example, a random substitution of p = 5% meant that 5% of A's were replaced with C's and 5% of T's with G's in randomly chosen locations of the sequence. It appears that the overall similarity of the sequence with the majority profile decreased along the upper portion of the sigmoid curve with increasing amounts of random substitution. c. Sigmoid curve of the stretch in panel (a) after subjecting it to a random substitution of A's and T's with C's and G's in sliding, random amounts p within a indicated ranges of $0<p<80\%$ and $-80\%<p<0$. (The latter case meant a reverse substitution of C's and G's with equal numbers of A's and T's.) The resulting curve reproduced the sigmoid curve quite accurately.

AT-content correlation plot will resemble one of the point clusters of Figure 13b. If we now plot the profile vs. the AT-content of the entire stretch the result will be a superimposition of the points in Figure 13a and b. There will be no breaks between them as in the simulation, because the plot includes also the transition sequences between flats and bursts. As a result the plot will appear as the upper portion of the sigmoid curve. Similarly, one can explain the appearance of the lower parts by assuming that very large changes of the AT-content in a burst.

Indeed, the absence of breaks in the plot could be demonstrated by generating transitions between flats and bursts in the following way. During the simulation the computer was allowed to chose the percentage of change p in a sliding, yet random way within a selected range of 0 to 80% and 0 to -80%. The resulting curve reproduced the sigmoid curve of mammals quite accurately (Figure13c).

A Possible Role of Transpositions in the Generation of the Sigmoid Curve

Based on the results of the above simulation one may attempt to explain the sigmoid curve by of one or several mechanisms that replaces a number z of A's or T's with the same number of C's or G's *on the same strand*. The term 'replace' could mean a conversion of z A's into G's and z T's into C's. It could also mean that the suspected mechanism deletes z A's and T's and inserts the same number of G's and C's in their place.

What could be the mechanism? It seems difficult to imagine a deletion-insertion mechanism or an enzymatic base modification mechanism that could achieve such peculiarly balanced replacements or conversions on the same strand. Therefore, I prefer to think of this unknown mechanism as the result of two distinct mechanisms: One mechanism converts (replaces) A's to G's, and a second generates the matching T-to-C conversions on the same strand.

Mechanism 1: A-To-G Conversion (Replacement)

This mechanism is assumed to turn a large number of A's into G's in a local restricted genome area of 10 – 20 Mb length. At certain times in evolution it converts (replaces) as many as 10 - 50 % of A's into G's in these limited stretches of the genome.

There are precedents for A-to-G conversions in the literature, although they occurred usually as single point mutations and often only in RNA's [28]. In addition I found that four kinds of point mutations, among them the A-to-G mutations are much more frequent than all others [29].

Mechanism 2: Matching of A-to-G And T-to-C Conversions (Replacements) on the Same Strand

Next we have to explain how an equal number of T's on the same strand segment were converted to C's, after the A-to-G conversion (replacement) had occurred on a strand segment. In this case, we may not have to look far. In fact, the very question implies that inversions/transpositions are involved. After all, the A-to-G conversion (replacement) without a matching T-to-C conversion must have created temporarily local strand asymmetries. Yet, I found that the strand symmetry signatures always remained flat in the area of strong bursts (Figure 3a). Therefore, one may suspect that inversions/transpositions followed the burst-creating processes and restored the strand symmetry in the region. As will be shown below, the inversions/transpositions that restored locally the strand symmetry were all that is needed to explain the required matching of A-to-G and T-to-C conversions (replacements) on the same strand.

Assume that initially the strand segment in question fulfilled all the requirements of genomes of the majority group, i.e. it had the majority profile and, therefore, it fulfilled Chargaff's second parity rule i.e. A = T=: x and C = G =: y on either strand.

	nucleotide:	A	T	C	G
Plus-strand:	numbers:	x	x	y	y
Minus-strand:	numbers:	x	x	y	y

Now let the unknown burst-creating mechanism replace a substantial number z of the A's on the Plus-strand to G's. Subsequently, the replication mechanism of DNA must replace the formerly matching T's on the Minus-strand with C's to match the new G's. Hence, as a result of base pairing the numbers of nucleotides on the strands become

	nucleotide:	A	T	C	G
Plus-strand:	numbers:	x-z	x	y	y+z
Minus-strand:	numbers:	x	x-z	y+z	y

Obviously, now the strands are no longer symmetrical and violate Chargaff's second parity rule. In addition their triplet differs substantially from the majority profile.

At this point we assume that a large number of inversions/transpositions occur in the same area. They will restore the strand symmetry [1] and thus yield according to equ. 4

	nucleotide:	A	T	C	G
Plus-strand:	numbers:	x-z/2	x-z/2	y+z/2	y+z/2
Minus-strand:	numbers:	x-z/2	x-z/2	y+z/2	y+z/2

Afterwards, this particular region has lost AT content and gained GC content by the same amount of z/2 on the same strand as is required by our analysis of the sigmoid curve. The strand symmetry is restored and with it the strand symmetry signature in this region is flat again.

By contrast, the action of the inversions/transpositions did not restore the majority profile in the local area of the segment. After all, the AT content is now much lower than the AT content of the genomes of the majority group. Nevertheless, the inversions/transpositions were able to enforce that the local change of the AT content follows the sigmoid curve accurately, as is shown in the example of Figure 14.

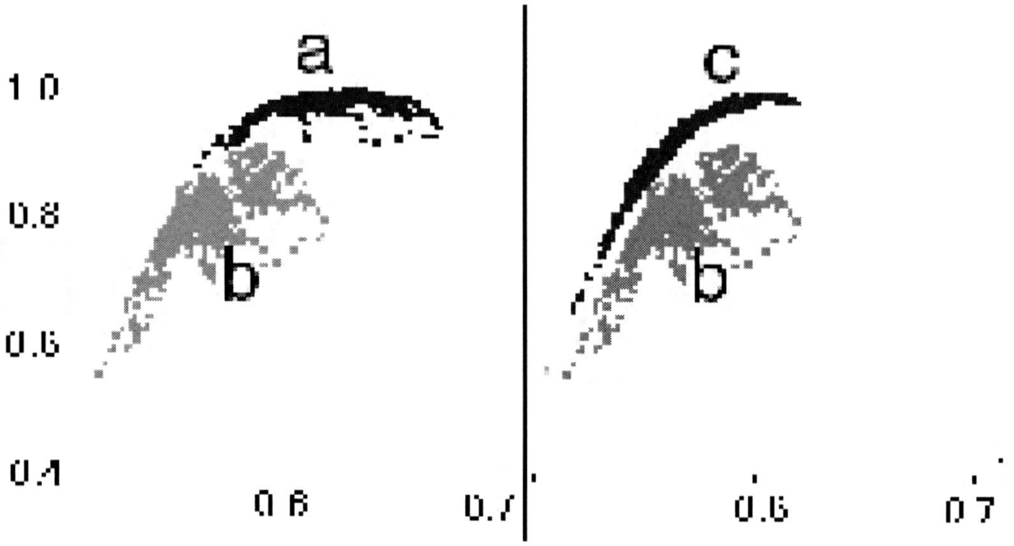

Figure 14. Effect of inversions/transpositions on the formation of the sigmoid curve. (Abscissa: AT-content measured in probe segments of 100 kb every 10 kb along a genome; Ordinate: Similarity between triplet profile and the majority profile of the same segments expressed as correlation coefficient between them). The Figure shows the sigmoid curve in the various stages of the described mechanism. Label 'a' (black dots): Sigmoid curve of the same 8 Mb large stretch of chimpanzee chromosome 14 that was used in the simulation of Figure 13. Label 'b' (gray dots): After a random A-to-G conversion in this stretch the average AT content is lowered to 55%, and the similarity with the majority profile is reduced. Label 'c' (black points): After exposing the stretch subsequently to 20,000 randomly placed inverted transpositions of a 5 kb size the data points fall again on the sigmoid curve, however they are shifted to a steeper part of it corresponding to their lower AT content.

In this example, we use as the initial segment the same 8 Mb large stretch of chimpanzee chromosome 14 that was used in the simulation of Figure 13. It shows the sigmoid curve of a flat part of the profile signature of this chromosome (Figure 14; label 'a'). As seen in the figure the segment has an average AT content of 62 %, corresponding to an A-content of 31%. After a computer-generated, random A-to-G conversion in this stretch the average A content was lowered to 26%, and the AT content was lowered to 57%. Thus the similarity with the majority profile was reduced, and the data points were no longer part of the sigmoid curve (Figure 14; label 'b'). However, after 20,000 computer-generated, randomly placed inverted transpositions of 5 kb size the data points fell again on the sigmoid curve, however they were shifted to a steeper part of it corresponding to their lower AT content. (Figure 14; label 'c'). Obviously, the inversions/ transpositions not only restored the strand symmetry (data not shown) but also enforced the relationship between local AT content and deviation from the majority profile that is described by the universal sigmoid curve.

Conclusion

The Universality of the Sigmoid Curve as a Function of Inversions/Transpositions

The sigmoid curve describes the local similarity between a genome's triplet profile and the majority profile as a function of the local AT-content. It has three remarkable features, namely its universality among the genomes tested, its sigmoid shape and its apparent dependence on only one variable. As will be shown in the Appendix , the first and second feature are not surprising if the third is true.

Therefore, in view of the astronomically large numbers of different genomes and triplet profiles, we need to understand how the similarity of a genome's triplet profile with the majority profile can depend on only one variable, namely the local AT-content of the genome sequence.

One would expect that it depends on at least three variables, namely the local A, T, and C contents (The G-content as the fourth variable is bound by the normalization requirement, i.e. it must add up to 100% together with the other three). On the other hand, it could not depend on more than 63 independent variables, because each test profile contains no more than 64 different frequencies f(Δ), one for each of the 64 triplets Δ, while one is bound by the normalization which requires that their sum has to add up to 100%

As described in the Results, this reduction of the conceivable variables to essentially a single one can be explained as the work of inversions/transpositions. Therefore, I suggest to add the universal sigmoid curve to the list of genome universals created by inversions/transpositions over evolutionary times.

The Creation of New Bursts During the Course of Evolution

Profile signatures appear to be markedly different for members of different kingdoms, phyla, and classes. Even members of the same class of *mammalia* such as platypus, rat, and chimpanzee have radically different profile signatures. In contrast, the profile signatures of members of the same order of primates such as humans, chimpanzee and rhesus ranged between 'similar' and 'practically identical'.

Based on my previous articles [1, 3], I had expected that the actions of countless inversions/transpositions during the past one billion years of evolution would have successfully erased all local deviations from the majority profile and the strand symmetry, and thus have turned both types of signatures into flat curves. This expectation was met to a reasonable degree by the strand symmetry signatures but clearly not by the profile signatures. Therefore, it seems that the effects of inversions/transpositions to level out any deviations from the majority profile were apparently much slower than the mechanisms that led to the emergence of new species.

At least, I had expected the equalization of all profile signatures to be a monotonous function of evolutionary time. In other words, no evolutionary 'younger' species to should have more bursts in its profile signatures than the genomes of an evolutionary 'older' ancestor genome that had more time to erase them.

Even that lesser expectation was obviously wrong. For example, the profile signatures of the phylogenetically 'older' opossum or even zebra fish were almost flat, whereas the profile signature of the 'younger' human showed 5 times as many bursts. Obviously, they must have been created *de novo* at some time after mammals branched away from the common ancestors of the zebra fish and other vertebrates. To be sure, these novel bursts were not drastic enough to remove the corresponding genomes from the majority group, but they clearly involved an identifiable and substantial portion of the 'new' genomes.

Is the Creation of New Bursts Linked to the Appearance of New Orders?

Considering that humans, chimpanzees and rhesus monkeys had very similar profile signatures, one may speculate, that their common ancestor, i.e. the first primate had this or a very similar profile signature as well. Likewise, one may speculate that the profile signatures of rat and mouse also shared their profile signatures with the first rodent. Yet, obviously, the profile signatures of rodents and primates were quite different from each other. Both, in turn, differed substantially from the profile signatures of the opossum and the platypus which belong to the yet different and much older orders of monotremes and marsupials within the class of mammals. It may seem, therefore, that new orders begin with new profile signatures that are essentially conserved as new members of the order evolve.

It remains to be seen whether going up the phylogenetic tree one encounters, indeed, new bursts and new profile signatures at every branching point of a new order. Our limited number of sequenced genomes does not permit to test this conjecture exhaustively at this time. For the time being one can only speculate that profile signatures may be markers of evolutionary branching points.

Of course, I am not suggesting that the creation of bursts was causal in the origin of a new order. Nevertheless, for the sake of illustrating the idea that inversions/transpositions could conceivably be instrumental in the creation of new orders, I shall try to bring the suggested burst-creating mechanism into the picture.

Conjecture about a Mechanism to Create New Orders

Imagine that a massive A-to-G conversion occurred at some point during evolution caused by an unknown mechanism. (It could also have been a replacement of A's with G's, or an insertion of a large number of G's.) One may view this event as a temporary and local reversion of the general trend of evolution to turn the GC-rich genomes of ancient organism (e.g., archaebacters, hydrothermal vent bacteria, and others) to the AT-rich genomes of the more modern organisms. In this sense, the event could be considered as a temporary return to the early conditions under which the earliest genomes arose.

If this reduction of the AT contents involved the entire genome, the mutation was probably lethal. On the other hand, if it involved only a few local 5 - 10 Mb large stretches ('GC-flares'), some genomes may have been able to retain or rescue their viability. Still, they would have been forced to cope with this kind of genome shock and may have unleashed large numbers of inversions/transpositions as an attempt to rescue their viability as suggested by Barbara McClintock [17]. Of course, these inversions/transpositions themselves may have spread far beyond the area of the GC-flares, damaged even more genes and their controls, and thus may have caused additional lethal mutations.

Yet, in cases, where the inversions/transpositions remained essentially concentrated in the area of the GC-flares, they transformed and stabilized them into strand symmetrical bursts that followed the sigmoid curve as described earlier. Nevertheless, a small number of the inversions/transpositions may still have 'strayed' from the immediate area of the flares and altered the controls and genes of other parts of this and other chromosomes. As a result, some of the still functional genomes became massively, yet not lethally altered in many genes and their controls.

In this way these genome-traumatic events that led to the GC-flares may have initiated major, practically simultaneous changes in the genome. They created possibly a drastically changed mutant organism which, in turn, was subsequently subjected to the next higher level of selection, namely the test whether it was survivable in its environment.

As a visible sign of these traumatic events, the genome of the mutant featured a certain pattern of bursts. Perhaps, one may consider these bursts as some kind of 'scars' left over from the locally suffered AT-reductions after they had been 'healed' by a series of inversions/transpositions which restored the local strand symmetry.

However, in general, these inversions/transpositions were not able to restore the former high AT-content of the loci of the GC-flares. If a Plus-strand is GC-rich, then so is the Minus-strand due to base pairing. Therefore, the strand swapping and sequence-transporting effect of inversions/transpositions could not have brought sufficiently many A's and T's back into the area of a flare, and thus could not have lowered the local GC-content as long as the inversions/transpositions operated predominantly around the area of the burst.

Of course, one could think of special AT-rich transposons that inserted and multiplied in the areas of the GC-flares and thus restored a high AT-content. And, perhaps, such events occurred and repaired some of the bursts. However, this was obviously not the case in the large number of GC-rich bursts that were found in the signatures described here. Their restoration to a high AT content would have required and still requires today an entirely different mechanism. Judging by the fact that the evolution of the 'modern', AT-rich genomes from presumably earlier GC-rich genomes required hundreds of millions of years, this mechanism is likely to be very slow. Besides, there seems to be no need to speed up this kind of restoration because, in principle, GC-richness does not pose any threat for the viability of a genome. After all, there are many completely viable organisms with GC-rich genomes.

As a result the descendents of the mutant would have continued to feature GC-rich, yet strand symmetrical bursts in the same pattern as the initial mutant. In case, they evolved further while replicating and preserving these drastically changed features of the initial mutant, they effectively rendered him the ancestor of a new order.

The Unsolved Problems of the Heritability and Simultaneity of the Mutations

To be sure, all suggestions and conjectures about evolutionary mechanisms that are based on the assumption of large scale genome variations suffer from the same unsolved problems. And the above conjecture is no exception.

One problem is the heritability of the mutations. How did the mutations, rare as the viable ones must have been, end up in the germ lines of organisms with sexual reproduction? Usually germ line cells constitute such a small fraction of the cells of an organism that it seems likely that the described traumatic events mutated mostly somatic cells. Do germ line cells have a better ability to deal with traumatic mutations and therefore provide many more viable mutants in spite of their small numbers? Or is there an unknown mechanism of horizontal gene transfer between somatic cells and the germ line on the time scale of evolution? To date there is no answer to these questions.

Another problem concerns the simultaneity of these mutations. Did all of them have to occur 'at once'? How long a time span is meant by the term 'at once'? In terms of evolutionary time scales, it could mean one or one million reproductive cycles, depending on the complexity of the organism and the length of its reproductive cycle. Of course, even a million reproductive cycles may be a very short time span by evolutionary standards. At any rate, we do not know the answer, but assume that the time span involved was very short compared to the age of the ancestor and the age of the resulting new species.

The Unknown Function of Genome Universals

Since the early days of Barbara McClintock's almost thirty year long battle to convince the scientific community that transpositions existed, they have revealed more and more diverse and profound consequences [30]. As she suggested herself in her Nobel lecture [17] their most important function may be found in their ability to empower genomes to react to environmental impacts "for which they were unprepared". In other words, they would be major agents for the evolution of life on Earth.

It appears that they also generated several additional universal genome properties during evolutionary times. As pointed out in the Introduction, one may speculate that cells use them to validate the legitimacy of unknown genomes that they may encounter during fertilization, cell fusion, virus infection and others. This universality may even amount to the existence of 'genome formats'. Such common formats may render the genomes of different organisms compatible with each other, help them to recognize the functionality of other genomes or their parts, and facilitate information transfer between them. Thus they may have the general function of preparing genomes for evolutionary changes resulting from encounters with parts of other genomes.

Still, the biological function(s) of the genome universals remain unknown. Their very nature as universals seems to exclude functions that offer selective advantages. After all, if they are universals then each organism will express them as much as its Darwinian competitors, and thus the scales of their selective advantages remain balanced.

Of course, this argument presumes that each organism expresses them to the same degree. That is clearly not the case. As mentioned earlier, there are organisms whose genomes do not belong to the majority group. Likewise, the degree of strand symmetry is very high, yet it is variable between different chromosomes and species. Perhaps, the degree by which genomes express some of these universals has, indeed, an effect of their selective success. For example, the reproductive success of a sexually reproducing species may depend on how much it expresses with the mentioned 'genome format', as that compliance may influence the success rate of the fusion of male and female pro-nuclei.

Unfortunately, to date no experiments have been carried out that tested whether modifications of the mentioned genome universals have any effect on the reproduction, development and function of an organism, or its ability to be the donor or receptor of horizontal gene transfer, or any other basic functions of genomes. For example, what effects would it have on development and gene control, if one would alter a genome's non-coding sequences in such a way that stretches of 100 kb or larger [1] violate the universal strand symmetry. Or, are site-specific recombinations, gene duplications, or horizontal gene transfer between members of the majority group inhibited if the genome sequences in question have substantially altered triplet profiles? Would bursts in profile signatures that violate the sigmoid curve amount to lethal mutations? I hope that future experiments of that kind will give the answers.

Materials and Methods

Genome sequences of human, chimpanzee, rhesus, opossum, rat, mouse, fugu, C.elegans, C.Briggsea, and zebra fish were obtained from the UCSC site.

Genome sequence of Arabidopsis and maize were obtained from the TIGR site.

Genome sequence of platypus and ciona intestinalis (sea squirt) were obtained from the ENSEMBL site.

The individual mitochondrial genomes used in this study were listed in [2]. Their source was the organelle databases of the EMBL-European Bioinformatics Institute (http://www.ebi.ac.uk/genomes/organelle.html).

The analysis program, "dnaorg.exe", was written by G.A.-B. using Visual C++ (Microsoft , Redmond, WA). Before evaluating the genome data, it replaced all repetitive sequences that had been masked by 'N' or 'n' with a neutral character '0' except in the computations of the sigmoid curve. In the latter case, all N's and n's were replaced with the next available unmasked base, in order to avoid that the beginning and end of masked, repetitive sequences produced artificial spikes in the strand symmetry signatures that distorted the sigmoid curve.

Appendix

A Basic Formula for Profile Signatures

Let f_M represent the majority profile, and h the triplet profile of the genome to be tested. Then we can define δH as the deviation from the majority profile

$$\vartheta H := h - f_M. \qquad (1)$$

The profile signature S_p (=similarity between the profiles of the test genome and the majority group) is given by

$$S_p = \text{correlation}\,[f_M,h] \qquad (2)$$
$$= \text{correlation}\,[f_M,\vartheta H+f_M]$$

$$= \frac{\text{covar}[f_M,\vartheta H+f_M]}{\sqrt{\text{var}[(f_M]}\bullet\sqrt{\text{var}[\vartheta H+f_M]}}.$$

Straightforward calculations yield

$$\text{covar}[f_M,\vartheta H+f_M]=(1/2)\{\text{var}[\vartheta H+f_M]+\text{var}[f_M]-\text{var}[\vartheta H]\}. \qquad (3)$$

and

$$\text{var}[\vartheta H+f_M]=\text{var}[f_M]+\text{var}[\vartheta H]+\text{covar}[f_M,\vartheta H]. \qquad (4)$$

Inserting equ. 3 and 4 into 2 yields as a basic formula for profile signatures S_p

$$S_p = \frac{1 + P}{\sqrt{\{1 + 2P + Q\})}}, \text{ with} \qquad (5)$$

$$P = \text{covar}[f_M,\vartheta H]/\ \text{var}[f_M], \qquad (5a)$$

and

$$Q = \text{var}[\vartheta H]/\ \text{var}[f_M]. \qquad (5b)$$

Given a fixed majority profile f_M, the value of var[f_M] is constant. In this case the value of S_p is a function of the position along the test genome h and can be presented in 2-dimensional graphs like Figure3b.

In order to analyze signatures, one would have to invoke and study additional variables, namely those that cause the test profiles h to change along the genome sequence. As mentioned in the Conclusion, there may be as many as 63 independent such variables. The analysis would then require some kind of correlation plots between S_p and these variables. Obviously, such plots cannot be depicted in 2-dimensional graphs, except in special cases. One such case is discussed in the next section.

The Universal Sigmoid Curve, a Special Case of the Profile Signature of Normalized, Strand Symmetrical Profiles

Note that the above considerations did not invoke the normalization or expression of strand symmetry in the test profiles. Consider, however, the simplified case that a normalized test profile h depends only on the base frequencies. Since there is no danger of misinterpretation, in the following the I simplify the notation as $A:=p_h(A)$, $C:=p_h(C)$, $T:=p_h(T)$, and $G:=p_h(G)$. Then,

$$h = h(A,C,T,G). \quad (6)$$

Hence, if there are relatively small changes δA, δT, δC, δG of the base frequencies, the profile changes by

$$\vartheta H = \delta h = (\partial h/\partial A)\delta A+(\partial h/\partial C)\delta C+(\partial h/\partial T)\delta T+(\partial h/\partial G)\delta G. \quad (7)$$

Assume that a series of inversions/transpositions have rendered the changes of the test sequence symmetrical, i.e.

$$\delta A = \delta T, \text{ and } \delta C = \delta G; \quad (8)$$

Since, the test profile is normalized (A+C+T+G=1), it follows that

$$\delta A+\delta T+\delta C+\delta G=0 \quad (9)$$

and therefore, after inserting equ.8,

$$\delta T=\delta A, \quad (10)$$

$$\delta C=-\delta A,$$

$$\delta G=-\delta A,$$

and

$\delta AT=\delta(A+T)=2\delta A$.

Inserting equ.10 into equ.7 yields

$$\delta h = Z\,\delta AT, \text{ with} \tag{11}$$

$$Z = (1/2)\,\{(\partial h/\partial A)+(\partial h/\partial T)-(\partial h/\partial C)-(\partial h/\partial G)\} \tag{12}$$

Since f_M is constant ($\delta f_M = 0$), according to equ. 1 and 11

$$\delta H = \delta h = Z \bullet \delta AT \tag{13}$$

δAT can also be written as

$$\delta AT = AT - AT_M. \tag{14}$$

where AT and AT_M are the AT-contents of the test- and majority profiles, respectively.

In other words, if the normalized test profile h depends only on the base frequencies A, C, T, and G and is strand symmetrical, then the profile symmetry depends only on the differences in their AT-contents. Inserting equations 13 and 14 into 5 yields

$$S_p = \frac{1 + P' \bullet (AT - AT_M)}{\sqrt{\{1 + 2P' \bullet (AT - AT_M) + Q' \bullet (AT - AT_M)^2\}}}, \text{ with} \tag{15}$$

$$P' = \text{covar}[f_M, Z] / \text{var}[f_M], \tag{15a}$$

and

$$Q' = \text{var}[Z] / \text{var}[f_M]. \tag{15b}$$

Equation 15 is shown in Figure15. It is quite similar to the universal sigmoid curve shown in Figure 10(second panel). For an accurate fit it assumes that

$$\text{covar}[f_M, Z] = 3\,\text{var}[f_M], \text{ and } \text{var}[Z] = 10\text{var}[f_M]. \tag{16}$$

The value of for the AT-content of the majority profile $AT_M = 0.75$ was derived from the majority profile.

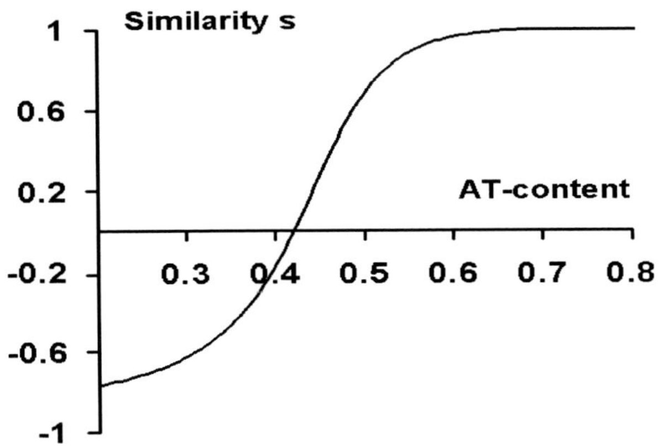

Figure 15. The profile signature S_p (equ.15) as a function of the AT content if both profiles are normalized and represent strand symmetrical sequences. The values of the variables used are AT_M = 0.75; P' = 3; and Q' = 10;

References

[1] Albrecht-Buehler G. (2006). Asymptotically increasing compliance of genomes with Chargaff's second parity rule through inversions and inverted transpositions. *Proc. Natl. Acad. Sci. USA* 103: 17828-17833

[2] 2. Albrecht-Buehler G. (2007). The three classes of triplet profiles of natural genomes. *Genomics* 89:596-601

[3] Albrecht-Buehler G. (2007). Inversions and inverted transpositions as the basis for an almost universal 'format' of genome sequences. *Genomics* 90:297-305

[4] Fickett, J.W., Torney, D.C., Wolf, D.R. (1992) Base compositional structure of genomes. *Genomics* 13(4): 1056-1064

[5] Baisnée, P.F., Hampson, St., Baldi, P. (2002) Why are complementary strands symmetric? Bioinformatics 188: 1021-1033.

[6] Prabhu, V.V. (1993) Symmetry observations in long nucleotide sequences. *Nucleic Acid Res*. 21: 2797-2800.

[7] Sanchez, J., Jose, M.V. (2002) Analysis of bilateral inverse symmetry in whole bacterial chromosomes. *Biochem. and Biophys. Res. Comm.* 299: 126-134.

[8] Mitchell, D., Bridge, R. (2006) A test of Chargaff's second rule. *Biochem. Biophys. Res. Comm.* 340:90-94

[9] Rudner, R., Karkas, J.D., Chargaff, E. (1968) Separation of B.subtilis DNA into complementary strands. *III Direct Analysis. Proc. Natl. Acad. Sci. USA* 603: 921-922.

[10] Rudner, R., Karkas, J.D., Chargaff, E. (1968) Separation of B.subtilis DNA into complementary strands. II. Template Functions and Composition as Determined by Transcription with RNA Polymerase. *Proc. Natl. Acad. Sci. USA* 603: 915-920.

[11] Forsdyke, D.R., Bell, S.J. (2004) Purine loading, stem-loops and Chargaff's second parity rule: a discussion of the application of elementary principles to early chemical observations. *Appl. Bioinformatics* 31: 3-8.

[12] Lobry, J.R. (1999) Evolution of DNA base composition under no-strand-bias conditions when the substitution rates are not constant. *J. Mol. Evol.* 166:719-723.

[13] Simons, C., Pheasant, M., Makunin, I.V., Mattick, J.S. (2006) Transposon-free regions in mammalian genomes. *Genome Res* .16: 164-172.

[14] Lander, E.S., Linton, L.M., Birren, B., Nusbaum, C., Zody, M.C., Baldwin, J., Devon, K., Dewar, K., Doyle, M., FitzHugh, W., et al. (2001) Initial sequencing and analysis of the human genome. *Nature* 409: 860–921.

[15] Gilbert, N., Labuda, D. (1999) CORE-SINEs: Eukaryotic short interspersed retroposing elements with common sequence motifs. Proc. Nat. Acad. Sci. USA 96: 2869-2874.

[16] Mighell, A.J., Markham, A.F., Robinson, P.A. (1997) Alu sequences. FEBS letters 417: 1-5

[17] McClintock , B. (1984) The significance of responses of the genome to challenge. *Science* 226: 792-801.

[18] Martin, S.L., Li, W.-L.P., Furano, A.V., Boissinot, S. (2005) The structures of mouse and human L1 elements reflect their insertion mechanism. Cytogenet. *Genome Res.* 110:223-228

[19] The RNA world. The Nature of Modern RNA Suggests a Prebiotic RNA world. 1993. eds. Gesteland RF, J F. Atkins. Cold Spring Harbor Monograph Series 24. Cold Spring Harbor Laboratory Press Cold Spring Harbor NY.

[20] Gilbert W. 1986. The RNA world. *Nature* 31: 9618

[21] Cech ThR. 1986. A model for the RNA-catalyzed replication of RNA. *Proc. Nat. Acad. Sci. USA* 83: 4360-4363.

[22] Guerrier-Takada C, Gardiner K, Marsh T, Pace N, Altman S.1983. The RNA moiety of ribonuclease P is the catalytic subunit of the enzyme. *Cell* 353: 849-57.

[23] Woese C. 1967.*The Genetic Code*. New York Harper and Row.

[24] Lindahl T. 1993. Instability and decay of the primary structure of DNA. *Nature (London)* 362, 709–715.

[25] Petrov DA, Hartl DL. 1999. Patterns of nucleotide substitution in *Drosophila* and mammalian genomes. *Proc. Natl. Acad. Sci. USA*. 96: 1475–1479.

[26] Cambareri EB, Jensen BC, Schabtach E, Selker EU. 1989. Repeat-induced G-C to A-T mutations in Neurospora. *Science* 244:1571-1575.

[27] Drobetsky EA, A. J. Grosovsky AJ, Glickman BW. 1987. The specificity of UV-induced mutations at an endogenous locus in mammalian cells. *Proc. Natl. Acad. Sci. USA* 84:9103-9107.

[28] Simpson L, Emeson RB. 1996. RNA editing. *Ann. Rev. Neuroscience* 19:27-52

[29] Albrecht-Buehler, G. (2007) The natural spectrum of point mutations as derived from the 'almost'-periodic sequences of genomes. (in preparation).

[30] Pennisi E. (2007) Jumping genes hop into the evolutionary limelight. *Sciene* 317:894-895

[31] Britten, RJ. (2006) Almost all human genes resulted from ancient duplication. *Proc.Nat. Acad. Sci. USA* 103:19027-19032

[32] Picard G, L'Heritier P. (1971). A maternally inherited factor inducing sterility in *Drosophila melanogaster. Drosoph. Inf. Serv.,* 46: 54-54.

In: DNA Transposable Elements Research
Editors: K. Yoshida, M. Aoki, pp. 109-131
ISBN: 978-1-60456-531-7
© 2008 Nova Science Publishers, Inc.

Chapter IV

Class 1 Integrons and Dissemination of Antibiotic Resistance Genes Among *Salmonellae*: From PCR- to DNA Microarray-Based Approach

***Tomas Majtan*[1*] *and Viktor Majtan*[2]**
[1]Department of Genomics and Biotechnology, Institute of Molecular Biology SAS, Dubravska cesta 21, Bratislava 45, 84551 Slovakia
[2]Department of Microbiology, Slovak Medical University, Limbova 14, Bratislava 37, 83303 Slovakia

Abstract

Apart from well-known bacterial methods of mutation, the horizontal transfer of genetic material within and between microbial genera plays crucial role in the emergence of novel antibiotic resistance genes and their dissemination observed worldwide. Rapid and widespread appearance of multidrug resistance and similar resistance patterns observed in closely as well as distantly related bacterial species are due to presence and activity of mobile genetic elements, especially integrons. Integrons are able to capture and exchange genes in a small mobile elements called cassettes in a process of site-specific recombination. They have been assigned to at least four distinct classes based on *int* integrase gene homology. Class 1 integrons are most prevalent in clinical isolates and usually associated with multidrug resistance phenotypes. This chapter describes PCR-based detection method of class 1 integrons and characterization of captured gene cassettes in clinical isolates of different serovars of *Salmonella enterica* in Slovakia during 2002-2006. In order to gain complex picture, we developed new microarray-based method for molecular characterization of *Salmonella* isolates in terms of integrase and antimicrobial resistance genes detection. We developed procedure using DNA microarray for localization of integrons as well as embedded gene cassettes carried either on chromosomal DNA or plasmids thus providing valuable insight to dissemination process and epidemiology of emerging resistance genes.

* Corresponding author: Tel: +421-2-59307429, Fax: +421-2-59307416, email: tomas.majtan@savba.sk

Introduction

Since the introduction of first antimicrobial compounds, antimicrobial resistance has been one of the major public health threats. Each day bacteria shows us why they are able to survive sometimes very hostile environmental conditions. Their ability to face different environmental threats is impressive and many times relies on their adaptability to rearrange and to redistribute genetic information. Although mutagenesis plays an important role in the evolution of antimicrobial resistance, the pivotal factor in acceleration of bacterial resistance represents the acquisition of antibiotic resistance genes. Bacteria possess a number of tools to rebuild bacterial genome enabling the cell to add, loose or alter genetic information thus update genetic content whenever possible. Despite of the abundance of resistance phenotypes observed among bacteria, there are a limited number of mechanisms by which these resistance traits are acquired. Bacteria can update their genetic content via (i) transformation by incorporating free DNA segments into their chromosome, (ii) transduction by gene transposition following bacteriophage infection and (iii) conjugation by plasmids or conjugative transposons [1]. Horizontal transfer of genes not limited only to resistance genes involving cell-to-cell transfer itself has become noteworthy important after engagement of gene transfer from one DNA molecule to another by mobile genetic elements or transposable elements. They include insertion sequences, composite, complex and conjugative transposons, transposing bacteriophages, integrons and recently described ISCR (Insertion Sequence Common Region) elements [2].

Integrons are genetic elements which recognize and capture exogenous open reading frames (ORFs) called gene cassettes by site-specific recombination and ensure their correct expression [3]. Integrons are not self-replicating and also not self-transferable DNA elements, but they can be part located within transposons. In order to transfer from one cell to another, they must be carried by a bacteriophage or more often by a plasmid. Five classes of integrons carrying antibiotic resistance gene cassettes have been reported so far based on the homology of their integrases, specific enzymes ensuring site-specific integration of captured gene cassettes, and novel yet unclassified integrons are still described [4]. However, class 1 integron is far the most prevalent and well characterized integron linked with multi-drug resistance (MDR) against antibiotics and disinfectants in various microbial species including *Salmonella spp.*. Class 1 integron (Figure 1) usually consists of 5'-conserved segment (5'-CS), 3'-CS and gene cassette(s) in between. The 5'-CS region is conserved and composed of three key elements: integrase *intI* gene, primary recombination site *attI* and out-ward oriented promoter P_C that directs transcription of captured gene cassettes. The 3'-CS region includes a *qacEΔ1* and a *sul1* gene, however, this region can be variable in length and even be absent. The integron-captured gene cassettes share the specific structural characteristics and generally contain single gene and an imperfect inverted repeat at the 3' end of the gene called an *attC* site (or 59-bp element). In *Salmonella*, class 1 integrons can be located on so-called Salmonella Genomic Island 1 (SGI1). SGI1 is a large (43 kb) integrative mobilizable chromosomal element, on which antibiotic resistance genes are clustered, flanked by two class 1 integrons [5].

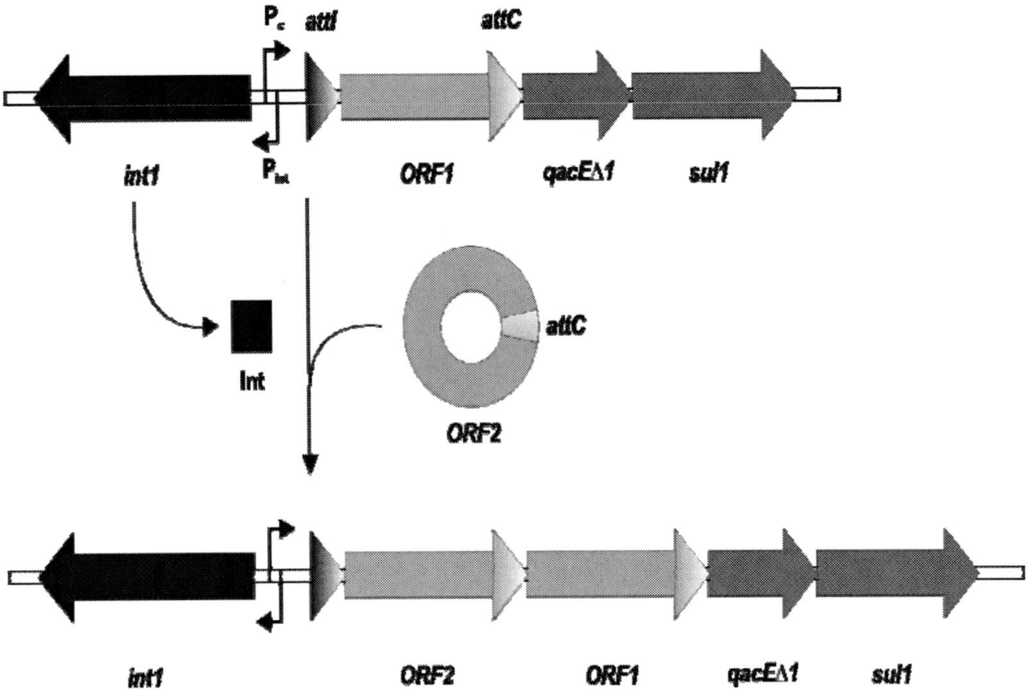

Figure 1. Class 1 integron structure and process of gene cassette capture. Class 1 integron usually consists of 5'-CS region (integrase gene *intI*, promoters P_{int} for integrase expression and P_c for captured gene cassettes expression and primary recombination site *attI*), 3'-CS region (typically includes *qacEΔ1* and a *sulI* genes) and embedded gene cassettes (designated as *ORF1* and *ORF2*). The circular gene cassette carrying *ORF2* is inserted at the specific recombination site *attI* by specific integrase Int downstream of the strong promoter P_c.

Detection of integrons and other mobile elements nowadays relies on specific set of PCRs. For instance, class 1 integrons are usually detected using PCR primers against 5'-CS and 3'-CS regions [6]. The presence of SGI1 is determined two pairs of PCR primers amplifying left and right SGI1 junction to the chromosome [7]. Resistance genes embedded in integrons or located elsewhere on chromosome or plasmids are often detected by resistance gene-specific PCRs [8, 9]. While antibiotic resistance is great emerging health issue, it is important to study the spread of antibiotic resistance genes to understand the relationship between resistance gene pool, its mobilization through transposons and integrons and dissemination among closely related as well as completely unrelated bacterial species. It is clear that traditional PCR approach is unable to encompass such a vast amount of genes and DNA elements. DNA microarray technique promises to detect thousands of genes simultaneously, thus providing complex insight into bacterial genomics. Hundreds of specific PCR or oligonucleotide probes attached to the microarray surface could provide after hybridization with labeled genetic sample complete genotype in single analysis. First microarrays detected only resistance genes [10], while recently developed DNA chips comprised additional genetic entities, such as integrons and transposons [11] as well as pathogenicity genes [9, 12].

The aim of the present study was to (i) determine the antimicrobial resistance profiles of *Salmonella* clinical isolates in Slovak Republic during 2002-2006, (ii) to characterize class 1 integrons in selected multi-drug resistant strains, (iii) to identify resistance markers conferring phenotype, (iv) to analyze the presence of specific integrases and resistance genes by DNA microarray system and finally (v) to test transmissibility of antimicrobial resistance markers as well as class 1 integrons via conjugation. We demonstrated the necessity and the usefulness of switch from PCR-based detection approach to more thorough molecular typing method, e.g. DNA microarray.

Material and Methods

Strains, Growth Conditions and DNA Isolation

A set of 15 *S. enterica* serovar Typhimurium and 3 *S. enterica* serovar Kentucky selected from the strain collection of the National Reference Centre for Phage Typing of Salmonellae (Bratislava, Slovakia) represented MDR clinical isolates from patients suffering from various forms of gastroenterocolitis during the years 2002–2006 (Table 1). The *S.* Typhimurium represents the most common serovar isolated from sporadic cases of human infections in Slovak Republic [13]. The *S.* Kentucky caused nosocomial outbreaks in two geographical regions of Slovakia [14]. All strains were biochemically tested by using the ENTEROTEST 16 (Lachema, Brno, Czech Republic) and were serotyped with commercial *Salmonella* somatic (O) and flagellar (H) antisera (Sanofi Diagnostic Pasteur, Marnes-la Coquette, France). Phage typing of *Salmonella* Typhimurium isolates was done according to the Anderson scheme [15]. The rifampicin-resistant *E. coli* K-12 3110 strain represent a recipient strain used for the conjugation assay.

Cells were grown on LB medium at 37°C. Genomic DNA was isolated by a DNeasy Tissue Mini kit (Qiagen, Germany). Plasmids were purified using a Large-construct kit (Qiagen).

Susceptibility Testing

Antimicrobial susceptibility was determined by the disc diffusion method according to the CLSI (formerly NCCLS) criteria on Mueller–Hinton agar plates [16]. The following antimicrobial agents with the respective quantities of active compound (in mg/disc) were tested: ampicillin 10, cefotaxime 30, ceftriaxone 30, chloramphenicol 30, ciprofloxacin 5, gentamicin 10, streptomycin 10, sulfisoxazole 250, tetracycline 30, trimethoprim 5 and a combination of trimethoprim/sulfamethoxazole 1.25/23.75. The *E. coli* ATCC 25922 was used as a control. The results were interpreted according to the CLSI criteria.

Table 1. Studied *Salmonella* strains and their characteristics

Strain	Phagetype	Resistance phenotype	Plasmid profile (kb)	Class 1 integrons (kb)
S. Typhimurium 10/02	DT193	AMP, CHL, CIP, GEN, STR, SUL, TET, TMP	90, 12, 6, 5	1.9 (*dfrA12-ORF-aadA2*)
S. Typhimurium 3/03	NT	GEN, STR, SUL, TET	10, 8, 6, 5.5, 2.1	1.0 (*aadA1*)
S. Typhimurium 8/03	NT	AMP, STR, SUL, SXT, TET, TMP	90	1.6 (*dfrA1-aadA1*)
S. Typhimurium 29/03	DT104	AMP, STR, SUL, TET	90	1.2 (*blaPSE-1*)
S. Typhimurium 40/03	DT104	AMP, CHL, STR, SUL, SXT, TET	90, 70, 35	1.0 (*aadA1*)
S. Typhimurium 19/04	DT104	AMP, CHL, STR, SUL, TET	90	1.2 (*blaPSE-1*), 1.0 (*aadA2*)
S. Typhimurium 6/05	NT	AMP, STR, SUL, SXT, TET, TMP	90, 50, 6.7, 4.3	1.6 (*dfrA14-aadA1*), 1.2 (*blaTEM-1*), 0.65 (*dfrA5*)
S. Typhimurium 14/05	DT104	AMP, CHL, STR, SUL, TET	90	1.2 (*blaPSE-1*), 1.0 (*aadA1*)
S. Typhimurium 36/05	DT120	AMP, CHL, STR, SUL, TET	90	1.2 (*blaPSE-1*), 1.0 (*aadA2*)
S. Typhimurium 55/05	DT20a	AMP, CHL, STR, SUL	90	1.2 (*blaTEM-1*), 1.0 (*aadA1*)
S. Typhimurium 163/05	ALR	CHL, STR, SUL, SXT, TET, TMP	30, 3.5	3.0 (*dfrB3-aadA1-catB2-aadA6*)
S. Typhimurium 19/06	NT	STR, SUL, TET	3	2.0 (*sat1-aadA1*)
S. Typhimurium 32/06	U302	AMP, CHL, STR, SUL, TET	90, 6.3, 3.6	1.2 (*blaPSE-1*), 1.0 (*aadA2*)
S. Typhimurium 90/06	DT1	AMP, STR, SUL, SXT, TMP	-	1.8 (*dfrA1-aadA1*)
S. Typhimurium 191/06	DT104L	AMP, STR, SUL, SXT, TET, TMP	7.5	1.6 (*dfrA1-aadA1*), 1.2 (*blaPSE-1*)
S. Kentucky 3297		AMP, CIP, STR, SUL, TET	90, 3.6, 2.4	1.2 (*blaTEM-1*), 1.0 (*aadA2*)
S. Kentucky 173/05		AMP, CIP, GEN, STR, SUL, TET	-	1.6 (*aacA5-aadA7*)
S. Kentucky 11411		AMP, CHL, CIP, CRO, CTX, GEN, STR, SUL, TET	3.6, 2.4	1.6 (*aacA5-aadA7*)

Abbreviations: NT – not-typeable, ALR – atypic lytic reaction, AMP – ampicillin, CTX – cefotaxime, CRO – ceftriaxone,CHL – chloramphenicol, CIP – ciprofloxacin, GEN – gentamicin, STR – streptomycin, SUL – sulfisoxazole, TET – tetracycline, TMP – trimethoprim, SXT – trimethoprim+sulfamethoxazole

Plasmid Profiling

Plasmid DNA was isolated by following the alkaline lysis protocol [17] with the modification according to our previous work [18]. Samples were analyzed by electrophoresis in 1x TBE buffer in 0.7% agarose gel. Plasmid sizes were determined according to the reference *E. coli* V517 strain, *Salmonella* Typhimurium LT2 strain and supercoiled DNA ladder (Gibco BRL, Paisley, UK) by using Gel-Pro Analyzer v4.5 (MediaCybernetics, USA).

Class 1 Integron Detection and Sequence Analysis

The PCR detection of class 1 integron content was performed essentially as described by Majtan et al. [18] using PCR primers against 5'-CS and 3'-CS regions (Table 2) [6]. PCR amplification was carried out in 50 μl reaction mixture containing 0.5 mM each dNTP, int1F and int1R primers, 1.0 U of *Taq* DNA polymerase (NEB) and 10 ng of purified bacterial total DNA. The temperature profile was: 94°C denaturation for 1 min, followed by 35 cycles of 94°C for 45 s with 45 s annealing at 56°C, then a 2.5 min extension step at 72°C. The PCR products were excised after electrophoresis on a 1% agarose gel and subsequently purified using the QIAEX II kit (Qiagen). Sequencing was performed in both directions by the BigDye method on an ABI 3100-Avant Genetic Analyzer (Applied Biosystems, USA). Obtained sequences were combined using Vector NTI Advance 10 (Invitrogen, USA). The similarity search was carried out using the BLAST.

Antimicrobial Resistance Genes PCR Detection

The most common resistance genes found in *Salmonella spp.* were detected by PCRs using PCR primers shown in Table 2. PCRs were conducted on the UNO-II thermocycler (Biometra, Germany) in a total volume of 50 μl containing 0.5 mM each dNTP and an appropriate primer pair at a concentration of 0.5 mmol each, 1.0 U of *Taq* DNA polymerase (NEB) and 10 ng of template DNA. The cycling conditions were as follows: an initial denaturation at 96°C for 5 min, followed by 36 cycles at 94°C for 1 min, 56°C for 45 s and 72°C for 1 min and completed by a final extension for 5 min at 72°C. The quality of PCR amplicons was examined by electrophoresis on a 1.2% agarose gel along with GeneRuler 100 bp DNA Ladder (Fermentas, Germany) as a marker.

Table 2. PCR primers used in this study

gene(s)	Forward Primer (5'->3')	Reverse Primer (5'->3')	amplicon size (bp)
class 1 int	GGCATCCAAGCAGCAAGC	AAGCAGACTTGACCTGAT	variable
blaOXA1/30	TATCTACAGCAGCGCCAGTG	TGGTGATCGCATTTTTCTTG	755
blaOXA2/53	AAGAAGGCACGCTAGAACG	CTACCCACCAACCCATACG	600
blaPSE	AATGGCAATCAGCGCTTC	ATCGCATCATTTCGCTCTG	650
blaSHV	GCGTAGGCATGATAGAAATGG	GGCGATTTGCTGATTTCG	707
blaTEM	CCAGAAACGCTGGTGAAAG	ATACGGGAGGGCTTACCATC	695
blaCMY/LAT	GATGCAGGAGCAGGCTATTC	CTCGACACGGACAGGGTTAG	997
blaCTXM1	AAGCCCTGGAGAAAAGTTCG	TATTGCCCTTAAGCCACGTC	540
blaCTXM2	GCAGAGTGAAACGCAAAAGC	TATCATTGGTGGTGCCGTAG	491

gene(s)	Forward Primer (5'->3')	Reverse Primer (5'->3')	amplicon size (bp)
aac(6')Ib/A4	TGACCTTGCGATGCTCTATG	TTAGGCATCACTGCGTGTTC	508
aadA/1/2/7	TTGTACGGCTCCGCAGTG	CAAATGCGGGACAACGTAAG	485
aadB	CGGCAGATGAGCGAAATC	CGCAAGACCTCAACCTTTTC	460
aph(3')Ic/A7	GAGCCATATTCAACGGGAAAC	CGAGCATCAAATGAAACTGC	800
catA	ATCCCAATGGCATCGTAAAG	CCTGCCACTCATCGCAGTA	611
catB	CTGGCTACTATCATGGGCACT	ACGGCAAACTCGAGCCAATA	535
cmlA	CGGATTCTTCAGGCTTGTG	ACCCACTAGCCACATTGGAG	630
floR	TTCGACATCCTCGCTTCAC	CGAAGGCAAAGCTGAATCC	700
dfrA1	GAATGGAGTTATCGGGAATGG	GGCGAAGTCTTGGGTAAAAA C	400
dfrA5/14	TGGTTGCGGTCCASACATAC	CGCCACCAGACACTATAACG	254
dfrA7/17	ATTTGACTCTATGGGTGTTCTTC C	TCAACGTGAACAGTAGACAA ATG	222
dfrA12/21	ATCGGGTTATTGGCAATGG	GGGAAGAAGGCGTCACC	352
sul1	ACCGAGGACTCCTTCTTCG	TCTAACCCTCGGTCTCTGG	800
sul2	TGGAGGCCGGTATCTGG	GTGCGGATGAAGTCAGCTC	684
qnrA	TAAAGTTTTTCAGCAAGAGGAT TTC	CCAAGGGTTCCAGCAGTTG	620

DNA Chip Construction

The presented DNA oligonucleotide chip was derived from our previously published DNA chip [9] by focusing on probes for detection of resistance genes and class 1 and 2 integrases. The recently developed spike-in mix [9] was included to monitor the labeling process and hybridization efficiency.

Lyophilized oligonucleotide probes (Table 3) were dissolved in 1x Micro Spotting Solution Plus (Telechem Int., USA) to a final concentration of 50 mmol. The DNA chips were prepared by using Pixsys 5500 microarrayer (Cartesian Technologies, USA) equipped with one ChipMaker CMP4 Micro Spotting Pin (Telechem Int.) at constant temperature (20°C) and humidity (42%). Oligonucleotide probes were deposited onto the SuperEpoxy slides (Telechem Int.) in four replicates per grid and two grids per slide, what allowed for two independent experiments at one time. Each grid thus consists of two identical subgrids, whose layout is shown in Figure 2A. The spotting result was verified using a CheckIt Chips Kit (Telechem Int.).

Table 3. DNA microarray oligonucleotide probes and their characteristics

gene/probe	sequence (5'->3')	length (bp)	Tm (°C)	GC (%)
aac(6')Ib/A4	CGAGCTCTGGTTGAGTTGCTGTTCAATGAT CCCGA	35	74.5	51.43

Table 3. (Continued)

gene/probe	sequence (5'->3')	length (bp)	Tm (°C)	GC (%)
aadA1/2/7	GGCGATGAGCGRAATGTAGTGCTTACKTT GTCCCG	35	N/A	54.29
aadB	GAGCTCGAGGCAATAGTTGAAATGCTCGG CGGG	33	75.2	57.58
aph(3')Ic/A7	TCCTGATGATGCATGGTTACTCACCACTGC GATCC	35	73.8	51.43
blaOXA1/30	TGGAACAGCAATCATACACCAAAGACGTG GATGCA	35	73.2	45.71
blaOXA2/53	ACGATAGTTGTGGCAGACGAACGCCAAGC GGATC	34	75.5	55.88
blaIMP	TGGCTTAATTCTCRATCTATYCCCACGTAT GCATC	35	N/A	40
blaPER	GTAGCAGTGTGGGGGCCTGACGATCTGGA ACCTT	34	74.4	58.82
blaPSE	TCATCCTAAGTGCTGTAGGTGGCCCCAAA GGCGTT	35	74.6	54.29
blaSHV	CCGGTGACGAACAGCTGGAGCGAAAGATC CACTAT	35	74.2	54.29
blaTEM	TCCCGGCAACAATTAATAGACTGGATGGA GGCGGA	35	75	51.43
blaVIM	TTGATTGATACAGCGTGGGGTGCGAAAAA CACAGC	35	74.4	48.57
blaGES	GGAGCGGTTTCTAGCATCGGGACACATGA CGGTT	34	74.8	55.88
blaCMY/LAT	GGCAATGACCAGACGCGTCCTGCAACCAT TAAAAC	35	74.6	51.43
blaCTXM1	CGAGAGCGATAAGCACCTGCTAAATCAGC GCGTTG	35	74.9	54.29
blaCTXM2	GACGTTTCGTCTGGATCGCACTGAACCTAC GCTGA	35	74.5	54.29
catA1	GCAAGATGTGGCGTGTTACGGTGAAAACC TGGCCT	35	75.6	54.29
catB3/8	GGTGATAGGYAGYCGCTCGTTGGTGACAA AAGATG	35	N/A	54.29
cmlA1/4/5/6/7	GGGATGCCTGATAGCTGGAGCAGTATTGC TTGCCA	35	74.7	54.29
floR	ATTGCGTCTCTGGGAGCAGCTTGGTCTTCA ACTGC	35	74.6	54.29
dfrA1	AATGGGAGCATTACCCAACCGAAAGTATG CGGTCG	35	74.4	51.43
dfrA5/14	GTGATTGGTTGCGGTCCASACATACCCTGG	33	N/A	57.58

	TCC			
dfrA7/17	TGGGTGTTCTTCCAAATCGMAAATATGCA GTAGTG	35	N/A	42.86
dfrA12/21	TACATCAAACCTTCGAGGGTGACGCCTTCT TCCCA	35	74.3	51.43
sul1	GAAGACGCGCTCGACGAGATTGTGCGGTT CTTC	33	75.2	57.58
sul2	GACATCATGGATCACATTGCGGCGTTCTTT GACG	34	74.6	50
ere(A)	TTCCGAAAAGCCTCTGATCGAATAGAGTC GATAGA	35	68.2	42.86
erm(B)	GTCTCGATTCAGCAATTGCTTAAGCTGCCA GCGGA	35	74.8	51.43
tet(A)	AATGATTGCCGACGGCACAGGCTACATCC TGCTT	34	75	52.94
tet(G)	TACACGATTATGGCATCAGCGCCGGTCTTA TGGGT	35	74.3	51.43
qacEΔ1	GGGACTCGGCGTCGTCATAATTACAGCCA TTGCCT	35	74.8	54.29
int1	AAGGCGAGCATCGTTTGTTCGCCCAGCTTC TGTAT	35	74.8	51.43
int2	CGCTGCTGTATGGTGCAGGTTTGCGCATTA ATGAA	35	74.8	48.57
DsRed	ATGGGAAAGGGTCATGAACTTTGAAGACG GTGGCG	35	75	51.43
dotACB	CCGCCGGAAGTGTCACAGGGGCTATCGTA AGCTTA	35	74.8	57.14
ORF02BFK2 0	GCCGTGGCAAGGTTCCGATATGTCTGGAT CCACTA	35	74.5	54.29
ORF12BFK2 0	TTCCGCTTTGTCATGCCTATCGACATCTTG GAGGG	35	74.5	51.43
ORF24BFK2 0	GCAACCAAGGTAACACGCTGCTCATGCAC TTGTCC	35	74.6	54.29
ORF41BFK2 0	TGGTCAAGTAGTTCGCGCCTATGCGCAAG CAAGAT	35	74.8	51.43
ORF47BFK2 0	GGTGTTCCGTGAGGGATTGAGCCAAGTAA CGACGT	35	74	54.29

Fluorescent Labeling

The total DNA and/or plasmids from *Salmonella* strains were fluorescently labeled via random priming [9]. Briefly, the 5 mg of template DNA was mixed with control spike-in mix and 9 mg of random primer (Invitrogen, USA). The mixture was subsequently heated for 10 min at 70°C and immediately chilled on ice. The DNA was labelled using 50 U of Klenow fragment exo⁻ (NEB, UK) in the presence of 2 mM dATP, dTTP, dGTP, 1 mM dCTP and 1 mM Cy5-dCTP (Amersham Biosciences, UK). The reaction was incubated at 37°C for 4 hours. Unreacted dye was removed using a QIAquick PCR Purification Kit (Qiagen). Labeling of plasmids followed essentially the same protocol except Cy5-dCTP was substituted with Cy3-dCTP. All dual channel microarray experiments were performed as dye swapped experiments.

Fluorescent labeling of class 1 integrons followed the same protocol as for their PCR detection. The specific PCR amplicon of class 1 integron excised from agarose gel served as template DNA for microarray target preparation. When strain contains more than one class 1 integron, purified PCR amplicons were mixed together in molar ratio(s) equal to 1:1 prior adding as a template for labeling. Fluorescently labeled nucleotide Cy5-dCTP in final concentration of 1 mM was included in dNTP mix. Unreacted dye was removed using a QIAquick PCR Purification Kit (Qiagen).

Hybridization and Post-Processing

Microarray slides were incubated in preheated (42°C) prehybridization buffer (5x SSC, 0.1% SDS and 1% BSA) for 1 h with an initial 15 min of vigorous shaking. The blocked slides were subsequently rinsed four times with milliQ water. Spotted DNA probes were heat denaturated by boiling milliQ water for 2 min and afterwards fixed by dipping the slides into pre-chilled 100% isopropanol for 5 s. Slides were dried with compressed air and incubated at 42°C until hybridization.

Hybridization was carried out in a Hybridization Chamber (Corning, USA). Vacuum concentrated fluorescently labeled DNA (either only Cy5-labeled genomic DNA, a mixture of Cy5-labeled genomic and Cy3-labeled plasmid DNA and vice versa for dye swap experiment or Cy5-labeled PCR amplicons of class 1 integrons) was completely dissolved in 15 μl of the 1x HybIt 2 hybridization buffer (Telechem Int.). Samples were subsequently denatured at 96°C for 5 min, immediately chilled on ice and spun in a microcentrifuge. Afterwards, they were incubated at 60°C water bath until applied onto the DNA chip. The hybridization was performed overnight at 42°C. Next day, the slides were washed using an initial wash in 2x SSC and 0.1% SDS preheated to 50°C for 5 min, followed by a 5 min wash in 1x SSC and finished by 1 min wash in 0.5x SSC both at room temperature. The slides were immediately dried with compressed air.

Normalization and Data Analysis

Processed DNA microarray slides were scanned by using a GeneTAC UC4 microarray scanner (Genomic Solutions, USA) at 10 mm resolution. Fluorescence signal intensities from hybridized probes were captured using Spotfinder by adaptive shape quantification followed by local background subtraction. The medians of the quadruplicate spots were further analyzed by using Microsoft Excel. Intensity median values from spike-in controls served for calculating the normalization factor, which was subsequently applied to adjust intensities from spots of interest. All spots with spike-in corrected intensity over three times signal from negative controls were considered as positive. Dual-channel experiments followed exactly the same procedure except that the intensities were additionally corrected according to ratio factor from spike-in controls and dye swap experiments.

Bacterial Conjugation

To determine the transmissibility of the resistance determinants and/or whole class 1 integrons, *Salmonella* donor strains were conjugated with recipient *E. coli* K-12 3110 as described previously [9]. Both donor and recipient strains were grown at 37°C to an OD_{550} of ~ 0.5. The recipient strain was thereafter grown for additional 15 min at 45°C prior to mating. The donor and recipient cells mixed in volume ratio 1:1 were conjugated for 60 min at 37°C and then pelleted by short spin in microcentrifuge. Cell pellet was resuspended in 200 μl of LB and immediately plated on the Tergitol-7 (HiMedia, India) agar plates supplemented with desired selective markers. Purified plasmid DNA from the obtained transconjugants was finally analysed using DNA microarray and PCRs for detection of antibiotic resistance genes as well as class 1 integrons.

Results

In the present study, 18 strains of two clinically relevant *Salmonella enterica* serotypes in Slovakia were studied in detail in terms of antimicrobial resistance, characterization of class 1 integrons and their ability to disseminate resistance genes among related as well as unrelated bacterial species. The 15 *S.* Typhimurium and 3 *S.* Kentucky strains were derived from a collection of 534 and 98, respectively, isolates obtained in 2002 – 2006 by the National Reference Centre for Phage Typing of Salmonellae (Bratislava, Slovakia). The most prevalent *S.* Typhimurium phagetype within selected strains was DT104, however, four isolates were untypeable according to the Anderson scheme and one provided atypical lytic reactions.

All selected *Salmonella* isolates were multi-drug resistant against 3 up to 9 tested antimicrobial substances (Table 1). Susceptibility testing revealed that all of selected strains were resistant to streptomycin and sulfisoxazole. Very commonly observed antimicrobial resistance was against tetracycline (89 %) as well as ampicillin (83 %). The all tested isolates except one were susceptible to the third generation of cephalosporins. The *S.* Kentucky 11411

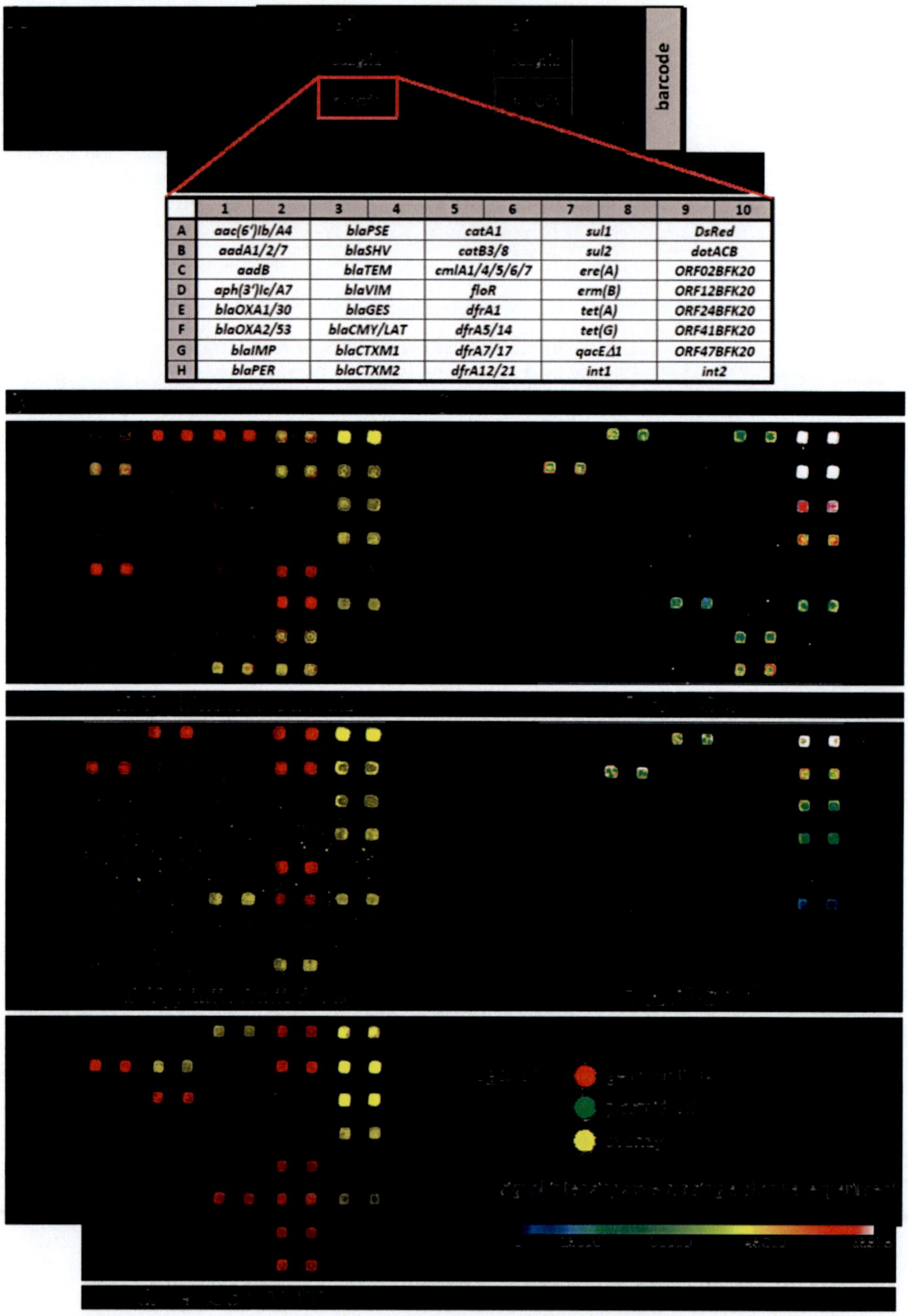

Figure 2. Oligonucleotide microarray for detection antibiotic resistance genes: chip layout (A) and hybridization scans (B, C). Chip layout (A) as well as hybridization scans of *Salmonella* strains (B) as well as *E. coli* transconjugants (C) showed one subgrid (duplicate spots) of two identical grids (each grid consists of two subgrids, i.e. quadruplicate spots) spotted on the surface of one slide. For dual-channel images, red palette represents genomic DNA, whereas green palette plasmid DNA. The yellow/orange color is designated for overlapping signals. For single channel image, rainbow palette was selected in order to show concentration-dependent signal intensities from spike-in controls as well as strong signal from monitored markers.

was found to be resistant against cefotaxime and ceftriaxone. Resistance against fluoroquinolone ciprofloxacin was detected in all three *S.* Kentucky strains, but only in one *S.* Typhimurium isolate.

The Table 1 shows the distribution of plasmids among selected multi-drug resistance *S.* Typhimurium and *S.* Kentucky clinical isolates. Only two strains, one from each serotype, did not possess any plasmid. The rest of strains carried one to five plasmids with molecular weight ranging from 2.1 kb up to 90 kb. The 90 kb serotype Typhimurium-specific virulence plasmid was found in 67 % of studied strains alone or along with the additional smaller plasmids.

PCR detection of class 1 integrons revealed the presence of at least one class 1 integron within each tested strain. The length of class 1 integrons PCR amplicons using primers against 5'-CS and 3'-CS regions was ranging from 0.65 kb in *S.* Typhimurium 6/05 up to 3.0 kb in *S.* Typhimurium 163/05. However, the most common length of PCR amplicons was 1.0 and 1.2, respectively (Figure 3). All detected class 1 integrons were PCR amplified, purified and sequenced in order to find out, which resistance gene cassettes are embedded within integrons. The genetic characteristics of class 1 integrons are shown in Table 1. The 27 sequenced class 1 integrons contained total of 39 captured resistance gene cassettes. The most prevalent resistance gene captured by class 1 integrons was the *aadA1* gene cassette conferring resistance against streptomycin. The 3 different homologs of *aadA1* gene cassette *aadA2*, *aadA6* and *aadA7* were also found within class 1 integrons. The *aadA7* cassette was found in 1.6 kb PCR amplicon exclusively along with *aacA5* gene cassette conferring resistance against other aminoglycoside gentamicin. The very common resistance against ampicillin was found due to the presence of beta-lactamase *blaPSE-1* in 6 cases or *blaTEM-1* in 3 cases. The resistance against trimethoprim was predominantly caused by the presence of *dfrA*: 4 homologs of *dfrA* (*dfrA1, dfrA5, dfrA12 and dfrA14*) were identified. In one case, resistance against trimethoprim was due to activity of *dfrB3* gene cassette. The *sat1* gene conferring resistance to streptothricin was found on class 1 integron in single *S.* Typhimurium 19/06 isolate as well as the *catB2* conferring resistance against chloramphenicol was found exclusively in *S.* Typhimurium 163/05. However, the 14 different classes of resistance gene cassettes were not able to completely determine the observed resistance phenotypes.

Twenty two genes implicated with antimicrobial resistance and conferring resistance against five families of antimicrobials, including aminoglycosides, amphenicols, beta-lactams, sulphonamides and trimethoprim had been chosen to determine their distribution among selected multi-drug resistant *Salmonella* clinical isolates and thus supplement the resistance genotypes obtained from class 1 integrons molecular characteristics. The presence of plasmid mediated resistance marker *qnrA* encoding resistance against nalidixic acid and fluoroquinolones was also monitored by PCR. The results of antimicrobial resistance genes PCR screening are shown in Table 4. Results indicate high degree of correlation between antimicrobial susceptibility testing and determined resistance genotypes, whereby multiple genes determined identical resistance phenotype. In example, resistance against aminoglycoside streptomycin had been found encoded by 4 different aminoglycoside adenylyl transferases *aadA1*, *aadA2*, *aadA6* or *aadA7*. Moreover, in several strains multiple genes conferring the same resistance phenotype. For instance, in *S.* Typhimurium 19/06 *sat1* and

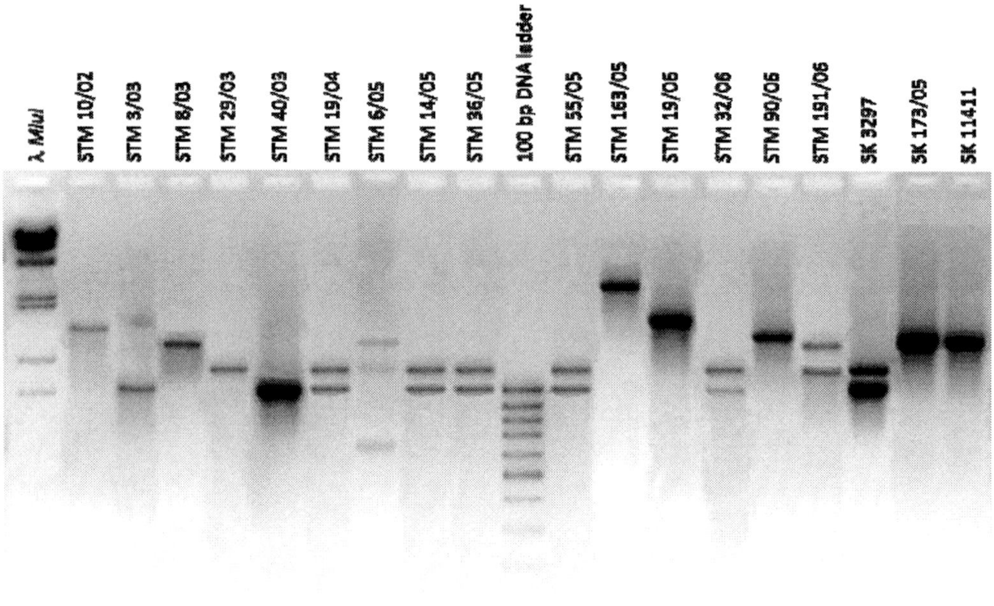

Figure 3. Class 1 integrons from *Salmonella spp.* strains. PCR amplicons generated with 5'-CS (int1F) and 3'-CS (int1R) primers.

aadA1 conferred resistance against streptomycin as well as in *S.* Kentucky 173/05 and 11411, respectively, resistance genes *aacA5* and *aadA7* coded for the same resistance phenotype against streptomycin. The resistance genes *dfrB3* and *catB2* carried on *S.* Typhimurium 163/05 class 1 integron were not detected by resistance genes PCR, while these alleles of dihydrofolate reductase and chloramphenicol acetyltransferase, respectively, differ substantially in nucleotide sequence of PCR primers binding sites from resistance genes, which PCR primers were designed for. On the other hand, in one *S.* Typhimurium and two *S.* Kentucky isolates we determined the presence of additional resistance genes to observed phenotype: *dfrA1* and *catA1* in *S.* Typhimurium 29/03, *dfrA5* in *S.* Kentucky 3297 and *dfrA14* in *S.* Kentucky 11411. We did not detect the presence of low level quinolone resistance determinant *qnrA* in any of the tested strains.

Table 4. Strain genotypes as determined by selected resistance genes PCRs

strain	resistance genes
S. Typhimurium 10/02	*aadA2*, blaOXA-1, catA1, dfrA12*, sul1, sul2*
S. Typhimurium 3/03	*aadA1*, sul2*
S. Typhimurium 8/03	*aadA1*, blaTEM-1, dfrA1*, sul1, sul2*
S. Typhimurium 29/03	*aadA1, blaPSE-1*, catA1, dfrA1, sul1, sul2*
S. Typhimurium 40/03	*aadA1*, blaPSE-1, floR, dfrA5, sul1*
S. Typhimurium 19/04	*aadA2*, blaPSE-1*, catA1, sul1*

Table 4 – Continued

strain	resistance genes
S. Typhimurium 6/05	*aadA1**, *blaTEM-1**, *dfrA5**, *dfrA14**, *sul1*, *sul2*
S. Typhimurium 14/05	*aadA1**, *blaPSE-1**, *floR*, *sul1*
S. Typhimurium 36/05	*aadA2**, *blaPSE-1**, *floR*, *sul1*
S. Typhimurium 55/05	*aadA1**, *blaTEM-1**, *floR*, *sul1*
S. Typhimurium 163/05	*aadA1**, *aadA6**, *sul1*, *sul2*
S. Typhimurium 19/06	*aadA1**, *sul1*
S. Typhimurium 32/06	*aadA2**, *blaPSE-1**, *floR*, *sul1*
S. Typhimurium 90/06	*aadA1**, *blaPSE-1*, *dfrA1**, *sul1*, *sul2*
S. Typhimurium 191/06	*aadA1**, *blaPSE-1**, *dfrA1**, *sul1*, *sul2*
S. Kentucky 3297	*aadA2**, *blaTEM-1**, *dfrA5*, *sul1*, *sul2*
S. Kentucky 173/05	*aadA7**, *blaTEM-1*, *sul1*
S. Kentucky 11411	*aadA7**, *blaSHV-2*, *blaTEM-1*, *catA1*, *dfrA14*, *sul1*, *sul2*

* - resistance gene found embedded in class 1 integron(s)

Previous results pointed out that it is highly desirable to monitor class 1 integrons together with resistance genes. In order to gain those results in one analysis, we developed and tested oligonucleotide chip for detection of class 1 and 2 integrases, key enzymes responsible for capturing gene cassettes by the corresponding integrons, along with 31 the most prevalent resistance genes among *Salmonella spp.* as well as whole Enterobacteriaceae family. The selection of oligonucleotide probes deposited at the microarray surface includes probes derived from genes encoding for resistance against tetracyclines, macrolides and disinfectants (based on quaternary ammonium compounds) in addition to five families of antimicrobials screened by resistance gene PCRs approach (Table 3). The distribution of monitored resistance genes across the genomes of *Salmonella* clinical isolates as determined by single channel DNA microarray analysis is shown in Table 5. For majority of probes, DNA microarray results verified resistance genes PCRs results as well as antimicrobial susceptibility analysis. However, a few discrepancies between identified genotypes and corresponding phenotypes as well as in genotype analysis by PCR versus DNA microarray had been detected. The chloramphenicol acetyltransferase *catA1* was detected by both PCR and DNA microarray analysis in *S.* Typhimurium 29/03, but strain failed to grow in growth medium supplemented with chloramphenicol. In latter case, the presence of *aadA1* was determined by PCR analysis in two streptomycin-resistant *S.* Typhimurium strains (*S.* Typhimurium 29/03 and 6/05), but DNA microarray analysis of whole genomic DNA did not support these findings. To demonstrate the DNA microarray possibilities and usefulness in genotyping of multi-drug resistant *Salmonella* strains, total genomic DNA and plasmids were differentially labeled and subsequently analyzed by dual channel DNA microarray. The captured images from such an analysis for selected *Salmonella* clinical isolates are shown in Figure 2B. The differential analysis of genomic and plasmid DNA of *S.* Typhimurium 10/02 revealed that class 1 integron carrying *dfrA12-ORFX-aadA2* gene cassettes is harbored on plasmid along with genes encoding for resistance against sulphonamides (*sul1*, *sul2*) and quaternary ammonium compounds (*qacEΔ1*). From three class 1 integrons detected in *S.*

Typhimurium 6/05 only the smallest one carrying *dfrA5* resistance gene cassette was found to be located on plasmid. On the other hand, no class 1 integron harbored on plasmid was found in *S.* Kentucky 11411. However, dual channel DNA microarray analysis of this strain had shown that resistance marker *blaSHV-2* coding for enhanced spectrum beta-lactamase (ESBL) conferring resistance against third generation of cephalosporins cefotaxime (CTX) and ceftriaxone (CRO) was carried on plasmids along with *catA1* gene. We have also tested the DNA microarray analysis of class 1 integrons (data not shown). Fluorescently labeled class 1 integron PCR amplicons were hybridized to presented chip in order to reveal the composition of class 1 integron embedded gene cassettes. This setup allowed for the resistance gene family membership determination of class 1 integrons gene cassettes without necessity of nucleotide sequence analysis.

Table 5. DNA microarray profiles of the 18 *Salmonella spp.* test strains and two *E. coli* transconjugants *

Probeset	Probe/Gene	*S.* Typhimurium 10/02	*S.* Typhimurium 3/03	*S.* Typhimurium 8/03	*S.* Typhimurium 29/03	*S.* Typhimurium 40/03	*S.* Typhimurium 19/04	*S.* Typhimurium 6/05	*S.* Typhimurium 14/05	*S.* Typhimurium 36/05	*S.* Typhimurium 55/05	*S.* Typhimurium 163/05	*S.* Typhimurium 19/06	*S.* Typhimurium 32/06	*S.* Typhimurium 90/06	*S.* Typhimurium 191/06	*S.* Kentucky 3297	*S.* Kentucky 173/05	*S.* Kentucky 11411	*E. coli* TC10.1	*E. coli* TC12.1
aminoglycosides	*aac(6')Ib/A4*																				
	aadA1/2/7	■	■	■	▒	■	■	▒	■	■	■	■	■	■	■	■	■	■	■	■	
	aadB																				
	aph(3')Ic/A7																				
β-lactams	*blaOXA1/30*	■																			
	blaOXA2/53																				
	blaIMP																				
	blaPER																				
	blaPSE	■			■	■	■		■	■				■	■	■				■	
	blaSHV																		■		■
	blaTEM			■				■			■						■	■	■		
	blaVIM																				
	blaGES																				
	blaCMY/LAT																				
	blaCTXM1																				
	blaCTXM2																				
amphenicols	*catA1*	■			▒		■												■		■
	catB3/8																				
	cmlA1/4/5/6/7																				
	floR					■			■	■	■			■							
sulphonamides	*dfrA1*			■	■										■	■					
	dfrA5/14					■		■									■		■	■	
	dfrA7/17																				

Table 5. (Continued)

Probeset	Probe/Gene	*S.* Typhimurium 10/02	*S.* Typhimurium 3/03	*S.* Typhimurium 8/03	*S.* Typhimurium 29/03	*S.* Typhimurium 40/03	*S.* Typhimurium 19/04	*S.* Typhimurium 6/05	*S.* Typhimurium 14/05	*S.* Typhimurium 36/05	*S.* Typhimurium 55/05	*S.* Typhimurium 163/05	*S.* Typhimurium 19/06	*S.* Typhimurium 32/06	*S.* Typhimurium 90/06	*S.* Typhimurium 191/06	*S.* Kentucky 3297	*S.* Kentucky 173/05	*S.* Kentucky 11411	*E. coli* TC10.1	*E. coli* TC12.1
	dfrA12/21	■	□	□	□	□	□	□	□	□	□	□	□	□	□	□	□	□	□	□	□
trimethoprim	*sul1*	■	□	■	■	■	■	■	■	■	■	■	■	■	■	■	■	■	■	■	□
	sul2	■	■	■	■	□	□	■	□	□	□	■	□	□	■	■	■	□	■	□	□
macrolides	*ere(A)*	□	□	□	□	□	□	□	□	□	□	□	□	□	□	□	□	□	□	□	□
	erm(B)	□	□	□	□	□	□	□	□	□	□	□	□	□	□	□	□	□	□	□	□
tetracyclines	*tet(A)*	■	□	□	□	□	■	■	■	■	□	■	■	□	□	■	□	■	■	□	□
	tet(G)	■	■	■	■	■	□	■	■	■	□	■	□	■	□	□	■	□	■	□	□
disinfectants	*qacEΔ1*	■	■	■	■	■	■	□	■	■	■	■	□	■	■	■	■	■	■	■	□
integrases	*intI*	■	■	■	■	■	■	■	■	■	■	■	■	■	■	■	■	■	■	■	□
	int2	□	□	□	□	□	□	□	□	□	□	□	□	□	□	□	□	□	□	□	□

* Positive results by both DNA microarray and PCR analysis compared with susceptibility testing are indicated by black filled blocks, whereas negative results are indicated by open white blocks. A white block with black diagonal lanes indicates target sequence detected by antimicrobial susceptibility testing and by PCR analysis, but was not confirmed by microarray analysis. A white block with black horizontal lanes indicates target sequence detected by DNA microarray and confirmed by PCR analysis, but does not correspond to susceptibility testing. The PCR validation was performed for all antimicrobial resistance genes except genetic markers conferring resistance against macrolides [*ere(A)*, *erm(B)*], tetracyclines [*tet(A)*, *tet(G)*] and ammonium compounds (*qacEΔ1*).

In order to determine horizontal transferability of resistance genes, we employed bacterial conjugation of selected *Salmonella* strains with rifampicin resistant *E. coli* K-12 3110 recipient strain followed by plasmid content analysis using DNA microarray approach. Plasmid(s) carrying *blaSHV-2* and *catA1* resistance determinants in *S.* Kentucky 11411 were transferred into recipient cells very efficiently (the efficiency of transfer was 78,3 ± 1.4 %) (transconjugant TC12.1, Figure 2C). More interesting result was observed from DNA microarray analysis of transconjugant TC10.1 (Figure 2C). The *dfrA5* and *intI* markers found carried on *S.* Typhimurium 6/05 plasmid(s) were detected along additional genes previously determined as chromosomally located: *aadA1*, *blaTEM-1*, *dfrA14*, *sulI* and *qacEΔ1*. However, efficiency of such conjugative transfer was lower than for *S.* Kentucky 11411 (32.7 ± 3.3 %). Resistance markers detected by DNA microarray were verified by resistance genes PCR analysis as well as class 1 integron PCR detection.

Conclusion

The main finding of this study was determination of high prevalence and diversity of class 1 integrons and resistance gene pool among *Salmonella* clinical isolates, introduction of antimicrobial resistance genotyping using DNA microarray and assessment of horizontal transferability of resistance genes as well as other genetic elements, such as class 1 integrons. According to the latest Enter-net Quarterly *Salmonella* Report [19], the most prevalent *Salmonella* serotypes represent Enteritidis (55.2 %) followed by Typhimurium (18.3 %) and Wirchow (1.7 %). Antimicrobial susceptibility testing results from 30 countries has shown that resistance against sulphonamides, tetracyclines, ampicillin, streptomycin and nalidixic acid is the most common among surveyed *Salmonella* strains (28.0 %, 18.3 %, 16.3 %, 15.5 % and 14.1 %, respectively). However, the most prevalent *Salmonella* serotype Enteritidis is also the most sensitive against tested antimicrobials. On the other hand, *Salmonella* serotypes Typhimurium and Kentucky expressed multidrug resistance profile against 4 or more antimicrobials in more than 45 % of tested isolates. It has been previously reported that multidrug resistance patterns in *Salmonella* Typhimurium and Kentucky are associated with the presence and activity of various genetic mobile elements, especially plasmids, class 1 integrons and SGI1 variants [5, 20, 21]. Class 1 integrons are unique: they can actively capture various gene cassettes, mainly resistance genes, and direct their transcription. The largest characterized class 1 integron was found on *E. coli* 160 kb conjugative plasmid and contained 9 functional resistance gene cassettes [22]. In our study, we found at least one class 1 integron in each *Salmonella* isolate with size up to 3.0 kb possessing 1 up to 4 resistance gene cassettes. The most typical class 1 integron PCR profile consisted of 2 PCR amplicons of 1.0 and 1.2 kb in size, respectively. The identical class 1 integron PCR profile has been described in many *S.* Typhimurium strains worldwide, especially in phagetype DT104, possessing *aadA2* gene conferring resistance against streptomycin and spectinomycin on 1.0 kb amplicon and *blaPSE-1* gene coding for ampicillin resistance located on 1.2 kb amplicon [23, 24]. The region, where are those class 1 integrons located, was latter sequenced and identified as *Salmonella* genomic island SGI1 [25]. The "original" SGI1 from *S.* Typhimurium DT104 possesses additional resistance genes *floR* and *tet(G)* flanked by class 1 integrons, what finally results in resistance phenotype against ampicillin, chloramphenicol, streptomycin, sulphonamides and tetracycline (ACSSuT pattern) often described as "pentaresistance". The SGI1 was subsequently found in various phagetypes of serovar Typhimurium: DT1, DT12, DT120 and U302, but it has been shown that DT12 and DT120 most likely arose from SGI1-possessing DT104 strain by changes in the susceptibility to the typing phages [26]. Furthermore, SGI1 has also been found in other *Salmonella* serovars from diverse sources worldwide, e.g. serotype Derby from swine in Japan [27] or serotype Newport from stools of a French patient suffering from gastroenteritis [28]. We can suggest that SGI1 may be able to spread among serovar Typhimurium phagetypes as well as other *Salmonella* serotypes via horizontal transfer. The SGI1 does not represent the stable and rigid structure and several SGI1 variants, designated as SGI1-A to SGI1-N, have been described in variety of *Salmonella* serotypes. The SGI1 variants differs to each other usually in antibiotic resistance gene cluster and were most likely generated by homologous recombination or by gene cassettes replacement/addition/deletion within class 1 integrons [29]. The class 1 integron with *aacA5-aadA7* aminoglycoside resistance gene array was found in two S. Kentucky 173/05 and 11411. The same cassettes array was identified in several *S.* Kentucky from Australia [30], Slovak Republic [14] and Ethiopia [31] and found always located on chromosome. Latter, this integron was found to be localized within SGI1-K variant along with an adjacent mercury resistance module [20]. We suggest that both *S.* Kentucky 173/05 and 11411 most likely carry the whole SGI1-K variant. Besides of SGI1 in *Salmonella* strains, the class 1 integrons can also be present elsewhere on chromosome or on plasmids. For instance, 1.9 kb class 1 integron with *dfrA12-ORF-aadA2* gene cassettes array from *S.* Typhimurium 10/02 was found on large plasmids in several *Salmonella* serovars such as Anatum, Braenderup and Saintpaul as well as 1.6 kb integron from *S.* Typhimurium 8/03 and 191/06 possessing *dfrA1-aadA1* resistance array was located on large plasmid in various *Salmonella* serotypes such as Braenderup and Rough [31]. The detection of class 1 integron in *S.* Typhimurium 163/05 carrying 4 resistance genes and presence of even larger integrons carrying up to 9 resistance genes [22] pointed out the existence of rich resistance gene cassettes pool, which is directly associated with the detection of multidrug resistance genotypes in hundreds and thousands of human or animal isolates. However, antimicrobial susceptibility testing does not always correspond to determined genotype as shown e.g. for *S.* Typhimurium 29/03. One of the reasons could be transcriptional control of resistance gene cassettes expression in class 1 integrons [32]. Transcription of usually promoterless gene cassettes driven by integron-borne Pc promoter directly depends on its strength. Other words it is mean, the farther the cassettes is from the promoter the less likely it will exhibit into a resistance phenotype. However, resistance cassettes constantly undergo rearrangement what can lead into sudden change in resistance phenotype. Although we found at least one class 1 integron in each tested clinical isolate, the detection of class 1 integrons by using 3'-CS and 5'-CS primers has a great limitation. It has been shown that some strains positive for intl integrase gene failed to yield a gene cassettes PCR amplicon(s) [33]. The reason of such failure is most likely in absence or sequence divergence of 3'-CS priming site [34]. We did not detect any of this case. However, we cannot exclude the presence of additional class 1 integrons with modified 3'-CS primer binding site. All these facts about transcriptional control of gene cassettes expression and possible misleading results from 5'-CS/3'-CS class 1 integron PCR detection pointed out the necessity of more complex approach in characterization of resistance genotype.

The PCRs targeting for 23 antimicrobial resistance genes revealed presence of additional resistance genes, which were most likely located on chromosome or carried by plasmids instead of embedded as resistance gene cassettes within class 1 integrons. The vast amount of genes conferring resistance to almost all antimicrobial groups commonly used in therapeutic praxis provides opportunity for their exchange and/or dissemination among closely related (i.e different *Salmonella* serovars or whole Enterobacteriaceae family) as well as evolutionary distant species. The hypothetical rate of such distribution is enhanced by their emplacement within mobile elements, such as class 1 integrons, in form of transferable resistance gene cassettes or their harboring on plasmids. Besides the high prevalence of resistance genes such as *aadA1*, *blaPSE-1*, *blaTEM-1*, *dfrA1* or *sul1* in *Salmonella* strains in present study as well as worldwide [5, 31], the most important finding was detection of *blaSHV-2* gene conferring resistance to the enhanced-spectrum beta-lactams (ESBL) such as cetriaxon and cefotaxim in

S. Kentucky 11411. Recently, we have described *blaSHV-2* resistance gene in different *S.* Kentucky strain 10944 and found it easily transferrable by bacterial conjugation into recipient *E. coli* strain [9]. We observed the same findings for *S.* Kentucky 11411. The facts that both strains were (i) isolated from hospitalized patients suffering from nosocomial infection, (ii) exhibited the similar resistance phenotype and (iii) were classified within the same plasmid profile as well as *XbaI*-PFGE profile group, pointed on the clonal dissemination of multi-drug resistant *S.* Kentucky strain across Slovak Republic [14]. Detection of *blaSHV* in *Salmonella* isolates is rare in comparison with other plasmid-borne ESBL *blaCTX-M* [35]. The resistance determinant against nalidixic acid and fluoroquinolones could be located either on choromosome or carried by plasmid usually along with other antimicrobial resistance genes [36]. Plasmid-mediated *qnrA* gene was not found in any of the tested strains. On the other hand, our previous study on *S.* Kentucky clinical isolates including *S.* Kentucky 11411 showed characteristic point mutations in quinolone resistance-determining region (QRDR) of DNA gyrase subunit gene *gyrA* (codon 83: TCC-Ser -> TTC-Phe; codon 87: GAC-Asp -> GGC-Gly), which conferred resistance against ciprofloxacin [14].

The complexity of relationship among antimicrobial resistance genes, mobile genetic elements (especially transposons, integrons, plasmids and particularly in *Salmonella* strains pathogenicity islands including SGI1), distribution and acquisition of resistance determinants and their dissemination (clonal as well as horizontal) requires complex detection approach. Here we presented a study employing DNA microarray for detection of resistance genes or gene cassettes carried on chromosome, plasmids or embedded in class 1 integrons in comparison to traditional techniques, such as PCR. Unlike studies reporting usefulness of DNA microarray technology in detection of resistance genes [12, 37, 38], our study has shown the potential of such analysis to answer the questions related to localization and transferability of DNA microarray screened genetic markers in addition to selective resistance genotyping. Careful selection and prudent combination of various fluorescently labeled targets (genomic DNA, plasmid DNA or integron PCR amplicons) was able to provide answers, which otherwise would be obtained by more laborious, time-consuming or expensive way (Southern blot, DNA sequencing,...). Our experimental design provided in single DNA microarray analysis information regarding resistance genotype of *Salmonella* isolate, presence of integron(s) and localization of resistance genes either on chromosome or plasmid. We have already demonstrated the usefulness of such complex approach using DNA microarray [9]. Although presented DNA microarray performed well, some discrepancies were identified in comparison to phenotype analysis or PCR determined genotype. These could arise from obvious reasons such as sequence divergence (point mutation(s)) of oligonucleotide probe to its respective labeled target or spatial and structural effects, especially for degenerated oligonucleotide probes. On the other hand, the great advantage of DNA microarray technology is its expandability for the new genetic markers of interest to deliver even more detailed characteristics of screened sample. The addition of plasmid replication and partitioning elements, prophage determinants and unique sequences of various transposons could be beneficial and have great impact for identification of the resistance genes distribution and dissemination routes. We feel that DNA microarray like in many other biological scientific areas can deliver new insight and open new horizons for detection of antimicrobial resistance genes, tracking of their transfer and dissemination and subsequent determination of epidemiological relationships.

Acknowledgement

We would like to thank Dr. Hana Drahovska (FNS CU, Bratislava, Slovakia) for help with the design of oligonucleotide probes and Ing. Marek Basler (ASCR, Prague, Czech Republic) for assistance with oligochip fabrication. This work was supported by APVV grant 21-052602 from the Slovak Research and Development Agency and partially from the National Reference Centre for Phage Typing of *Salmonellae* (Bratislava, Slovakia) core budget.

References

[1] Alekshun, M.N. and S.B. Levy, Molecular mechanisms of antibacterial multidrug resistance. *Cell*, 2007. 128(6): p. 1037-50.

[2] Toleman, M.A., P.M. Bennett, and T.R. Walsh, ISCR elements: novel gene-capturing systems of the 21st century? Microbiol. Mol. Biol. Rev., 2006. 70(2): p. 296-316.

[3] Mazel, D., Integrons: agents of bacterial evolutio*n. Nat. Rev. Microbiol.,* 2006. 4(8): p. 608-20.

[4] Elsaied, H., et al., *Novel and diverse integron integrase genes and integron-like gene cassettes are prevalent in deep-sea hydrothermal vents. Environ. Microbiol,* 2007. 9(9): p. 2298-312.

[5] Vo, A.T., et al., Antibiotic resistance, integrons and Salmonella genomic island 1 among non-typhoidal Salmonella serovars in The Netherlands. *Int. J. Antimicrob. Agents,* 2006. 28(3): p. 172-9.

[6] Levesque, C., et al., PCR mapping of integrons reveals several novel combinations of resistance genes. *Antimicrob. Agents Chemother.*, 1995. 39(1): p. 185-91.

[7] Doublet, B., et al., Variant Salmonella genomic island 1 antibiotic resistance gene cluster in Salmonella enterica serovar Albany. *Emerg. Infect. Dis.*, 2003. 9(5): p. 585-91.

[8] Miko, A., et al., Molecular mechanisms of resistance in multidrug-resistant serovars of Salmonella enterica isolated from foods in Germany. *J. Antimicrob. Chemother.,* 2005. 56(6): p. 1025-33.

[9] Majtan, T., et al., Oligonucleotide microarray for molecular characterization and genotyping of Salmonella spp. strains. *J. Antimicrob. Chemother.*, 2007. 60(5): p. 937-46.

[10] van Hoek, A.H., et al., Detection of antibiotic resistance genes in different Salmonella serovars by oligonucleotide microarray analysis. *J. Microbiol. Methods,* 2005. 62(1): p. 13-23.

[11] Frye, J.G., et al., DNA microarray detection of antimicrobial resistance genes in diverse bacteria. *Int. J. Antimicrob. Agents*, 2006. 27(2): p. 138-51.

[12] Malorny, B., et al., Molecular characterisation of Salmonella strains by an oligonucleotide multiprobe microarray. *Mol. Cell Probes,* 2007. 21(1): p. 56-65.

[13] Majtanova, L., [Occurrence of phage types of non-typhoid Salmonella serovar in the Slovak Republic 2000-2003]. *Epidemiol. Mikrobiol. Imunol.*, 2004. 53(2): p. 70-3.

[14] Majtan, V., et al., Salmonella enterica serovar Kentucky: antimicrobial resistance and molecular analysis of clinical isolates from the Slovak Republic. *Jpn. J. Infect. Dis.*, 2006. 59(6): p. 358-62.

[15] Anderson, E.S., et al., Bacteriophage-Typing Designations of Salmonella-Typhimurium. *Journal of Hygiene,* 1977. 78(2): p. 297-300.

[16] Institute, C.a.L.S., Performance Standards for Antimicrobial Disc Susceptibility Tests—Ninth Edition: Approved Standard M2-A9. 2006, CLSI: Wayne, PA, USA.

[17] Birnboim, H.C. and J. Doly, A rapid alkaline extraction procedure for screening recombinant plasmid DNA. *Nucleic. Acids Res.,* 1979. 7(6): p. 1513-23.

[18] Majtan, V., L. Majtanova, and L. Kovac, Analysis of integrons in human isolates of Salmonella enterica serovar typhimurium isolated in the Slovak Republic. *FEMS Microbiol Lett,* 2004. 239(1): p. 25-31.

[19] Gill, N. and J. Threlfall, Enter-net Quaterly Salmonella Report April-June 2007/2. 2007.

[20] Levings, R.S., et al., SGI1-K, a variant of the SGI1 genomic island carrying a mercury resistance region, in Salmonella enterica serovar Kentucky. *Antimicrob. Agents Chemother.*, 2007. 51(1): p. 317-23.

[21] Zaidi, M.B., et al., Rapid and widespread dissemination of multidrug-resistant blaCMY-2 Salmonella Typhimurium in Mexico. *J. Antimicrob. Chemother.*, 2007. 60(2): p. 398-401.

[22] Naas, T., et al., Characterization of In53, a class 1 plasmid- and composite transposon-located integron of Escherichia coli which carries an unusual array of gene cassettes. *J. Bacteriol.*, 2001. 183(1): p. 235-49.

[23] Ridley, A. and E.J. Threlfall, Molecular epidemiology of antibiotic resistance genes in multiresistant epidemic Salmonella typhimurium DT 104. *Microb. Drug Resist.*, 1998. 4(2): p. 113-8.

[24] Sandvang, D., F.M. Aarestrup, and L.B. Jensen, Characterisation of integrons and antibiotic resistance genes in Danish multiresistant Salmonella enterica Typhimurium DT104. *FEMS Microbiol. Lett.,* 1998. 160(1): p. 37-41.

[25] Boyd, D., et al., Complete nucleotide sequence of a 43-kilobase genomic island associated with the multidrug resistance region of Salmonella enterica serovar Typhimurium DT104 and its identification in phage type DT120 and serovar Agona. *J. Bacteriol.*, 2001. 183(19): p. 5725-32.

[26] Lawson, A.J., et al., Multiply resistant (MR) Salmonella enterica serotype Typhimurium DT 12 and DT 120: a case of MR DT 104 in disguise? *Emerg. Infect. Dis.*, 2002. 8(4): p. 434-6.

[27] Akiba, M., et al., Detection and characterization of variant Salmonella genomic island 1s from Salmonella Derby isolates. *Jpn. J. Infect. Dis.*, 2006. 59(5): p. 341-5.

[28] Cloeckaert, A., et al., Variant Salmonella genomic island 1-L antibiotic resistance gene cluster in Salmonella enterica serovar Newport. *Antimicrob. Agents Chemother.,* 2006. 50(11): p. 3944-6.

[29] Mulvey, M.R., et al., The genetics of Salmonella genomic island 1. *Microbes Infect,.* 2006. 8(7): p. 1915-22.

[30] Levings, R.S., et al., New integron-associated gene cassette encoding a 3-N-aminoglycoside acetyltransferase. Antimicrob. *Agents Chemother.*, 2005. 49(3): p. 1238-41.

[31] Molla, B., et al., Class 1 integrons and resistance gene cassettes among multidrug resistant Salmonella serovars isolated from slaughter animals and foods of animal origin in Ethiopia. *Acta Trop.*, 2007. 103(2): p. 142-9.

[32] Depardieu, F., et al., Modes and modulations of antibiotic resistance gene expression. *Clin. Microbiol. Rev.,* 2007. 20(1): p. 79-114.

[33] Hsu, S.C., et al., Characterisation of antimicrobial resistance patterns and class 1 integrons among Escherichia coli and Salmonella enterica serovar Choleraesuis strains isolated from humans and swine in Taiwan. *Int. J. Antimicrob. Agents*, 2006. 27(5): p. 383-91.

[34] Post, V., G.D. Recchia, and R.M. Hall, Detection of gene cassettes in Tn402-like class 1 integrons. *Antimicrob. Agents Chemother.*, 2007. 51(9): p. 3467-8.

[35] Li, X.Z., et al., beta-Lactam resistance and beta-lactamases in bacteria of animal origin. *Vet. Microbiol.,* 2007. 121(3-4): p. 197-214.

[36] Li, X.Z., Quinolone resistance in bacteria: emphasis on plasmid-mediated mechanisms. *Int. J. Antimicrob. Agents.*, 2005. 25(6): p. 453-63.

[37] Call, D.R., et al., Identifying antimicrobial resistance genes with DNA microarrays. *Antimicrob. Agents Chemother.*, 2003. 47(10): p. 3290-5.

[38] Volokhov, D., et al., Microarray analysis of erythromycin resistance determinants. *J. Appl. Microbiol.*, 2003. 95(4): p. 787-98.

In: DNA Transposable Elements Research
Editors: K. Yoshida, M. Aoki, pp. 133-155
ISBN: 978-1-60456-531-7
© 2008 Nova Science Publishers, Inc.

Chapter V

The Phenotypic Effects of Insertion Sequence-Mediated Gene Activation in Human Bacterial Pathogens

C. R. E. McEvoy, M. Newton-Foot, P. D. van Helden, and R. M. Warren
DST/NRF Centre of Excellence for Biomedical Tuberculosis Research / MRC Centre for Molecular and Cellular Biology, Division of Molecular Biology and Human Genetics, Faculty of Health Sciences, Stellenbosch University, Tygerberg 7505, South Africa

Abstract

Insertion sequences are regarded as the simplest mobile genetic elements, encoding only the information necessary to enable and regulate their transposition. Bacterial insertion sequences display high diversity and may be categorised into multiple families based on related transposase domains, open reading frame structures, terminal inverted repeats (IRs), and similarities in the outcome of the target site DNA upon integration. Although originally considered to be passive genetic elements or 'junk' DNA, many years of accumulated evidence has now conclusively demonstrated that they can act as important regulators of gene expression and mediators of genome plasticity. Thus, despite their structural simplicity, the genotypic and subsequent phenotypic effects of bacterial insertion sequences may be complex. This complexity is enhanced by the fact that many bacterial insertion sequences are located on extrachromosomal DNA elements that may be transmitted between species and have the capacity to integrate into host chromosomes, carrying accessory genes with them. Typically, the transposition of an insertion sequence results in a negative effect on gene expression in the target DNA due to the disruption of coding or promoter sequences. However, many insertion sequences have been shown to possess the ability, through several related but discrete mechanisms, to upregulate genes downstream from the insertion sequence integration site. Here we review some of the phenotypic consequences of insertion sequence-mediated gene activation, focussing on its effects in human bacterial pathogens. In particular we discuss how insertion sequence-mediated gene activation can alter bacterial virulence and lead to antibiotic resistance.

Introduction

Insertion sequences are small (generally < 2 kb), genetically compact elements that encode only for functions involved in their mobility. With some exceptions, insertion sequence length is defined by 10 – 40 bp terminal IRs that may be involved in transposase binding or in DNA cleavage and strand transfer reactions that result in transposition of the element. Upon genome integration most insertion sequences produce short (2 – 14 bp) sequences that are direct repeats of the target DNA flanking the element. Transposition may be either non-replicative, resulting in the relocation of the insertion sequence from one site to another (cut and paste), or replicative, where the element creates a new copy of itself which inserts elsewhere in the genome (copy and paste) (Figure 1) (Mahillon and Chandler, 1998). The target sequence specificity of insertion sequences varies from strict to random, although even those elements that display apparently random target DNA specificity generally reveal integration hot-spots at a genome-wide level (Warren et al., 2000). This suggests that more subtle larger-scale DNA structures may also influence integration site.

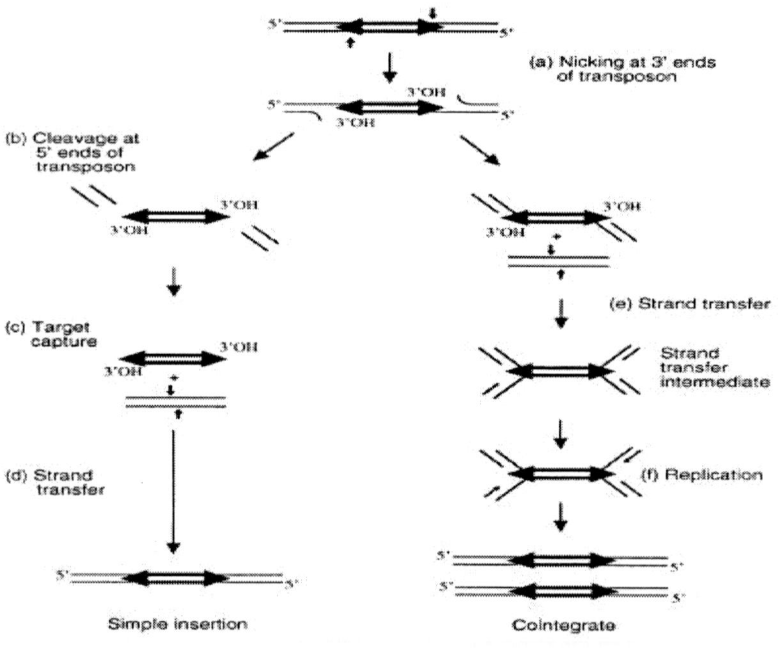

Figure 1. Summary of the biochemical steps that occur in non-replicative and replicative transposition pathways. Nicking at the 3′ ends is the initial step in both pathways (a). In the simple insertion pathway, this is followed by cleavage of the 5′-flanking DNA (b), which generates an excised transposon. Interaction with a target (c) allows strand transfer to occur, which results in a simple insertion (d). Replicative transposition (a, e and f) occurs via a strand transfer reaction (e) involving the nicked transposon and a target to generate a strand transfer intermediate. Replication of this intermediate using the exposed 3′ OHs in the target DNA as priming sites (f) results in duplication of the transposon and a co-integrate structure. IRs are represented by filled triangles, the transposon DNA is shown as a thick line, flanking and target DNAs as thin lines, and cleavage sites by small vertical arrows. Reproduced from Tavakoli and Derbyshire (2001) (Tavakoli and Derbyshire, 2001) with the kind permission of the authors.

Insertion sequences may be viewed as important agents in promoting and maintaining genome plasticity both within and between species (Frost et al., 2005). This, in turn, may provide the evolutionary tools required for the emergence of advantageous traits. This idea is strengthened by examples of extreme insertion sequence expansion (other less extreme examples also exist) that have been implicated in the emergence of pathogenicity in members of the Bordetellae and Yersinae bacterial families (Parkhill et al., 2003; Chain et al., 2004). However, despite examples such as these, considerable debate has focussed on the question of whether such 'selfish' DNA elements truly provide an overall benefit to their hosts since, although they may provide beneficial effects, these elements may also be viewed as extremely effective parasites that replicate themselves at the expense of their host (Doolittle and Sapienza, 1980; Orgel and Crick, 1980). This viewpoint supposes that insertion sequences persist within genomes simply because of their highly infectious character rather than any fitness advantage they provide. Recent results from the *in silico* analysis of 18 fully sequenced diverse bacterial genomes suggests that most insertion sequences are evolutionarily young and have been acquired relatively recently (Wagner, 2006). This suggests that although they might undergo periods of expansion and be transiently beneficial, they are detrimental to their hosts over extended evolutionary time periods, resulting in their eventual extinction in bacterial lineages.

Insertion sequences are widespread in eukaryotes, prokaryotes (including archaea), viruses and plasmids. Indeed, they may be viewed as an inevitable consequence of Darwinian selection on DNA sequences. Apart from undergoing transpositional events within individual organisms, they may also be transferred between genomes of sometimes widely dissimilar species by horizontal gene transfer (HGT) mechanisms. In bacteria, they may be located within the chromosome or on extrachromosomal elements and they are often found in association with composite transposons containing gene cassettes that are capable of conferring adaptive traits such as drug resistance and altered metabolic capabilities (Figure 2). Insertion sequences have been shown to be involved in the transfer of these traits between genomes or between genome and extrachromosomal element. Apart from their role in transferring relatively large genetic elements between genomes, they also play important roles within individual genomes due to their transpositional activity. The most obvious result of insertion sequence transposition is insertional mutagenesis, caused by the integration of the element into the coding or promoter region of a gene and resulting in a gene knockout phenotype. This is the basis for the widely used transposon mutagenesis technique, a valuable tool in gene function studies which is able to determine the phenotypic effects of specific gene knockouts in a range of organisms. Other genomic alterations caused by insertion sequences include duplications, deletions (often caused by recombination events between elements), and possibly integration induced point mutations and other micro-DNA rearrangements (Warren et al., 2000). It has also been known for many years that certain bacterial insertion sequences are capable of upregulating gene expression by means of their ability to act either as fully functional mobile promoters or by forming hybrid promoters with the DNA target sequence (Saedler et al., 1974; Pilacinski et al., 1977; Zafarullah et al., 1981). Insertion sequences may also upregulate gene expression by disrupting negative *cis* or *trans* regulators. Thus, they may have positive effects on gene expression as well as the more frequently encountered negative ones.

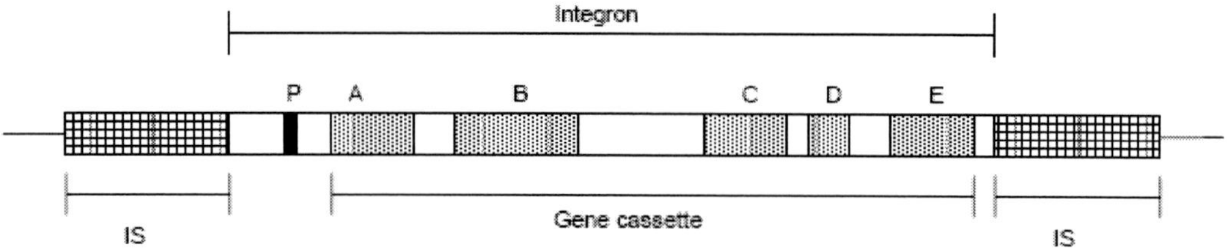

Figure 2. Bacterial Composite Transposon. Diagram of a hypothetical bacterial composite transposon flanked by two identical insertion sequences (IS, hatched region) and containing an integron comprising a promoter (P) and a gene cassette consisting of 5 genes A – E (shaded). Bacterial composite transposons are associated with identical or near-identical insertion sequences flanking a non-transposable DNA sequence. The non-transposable sequence often contains an integron, defined as a gene capture system where gene cassettes can be incorporated and expressed, exchanged or expelled. Gene cassettes comprise a series of genes used for a single function such as a novel biochemical pathway or drug resistance. The flanking insertion sequences allow mobilisation of the entire composite transposon between chromosomal and extra-chromosomal elements. Insertion sequences may also be present within the integron where they can exert additional influences on gene expression and mobilisation.

Pathogenic bacteria continue to exert an enormous social and economic cost on human society, particularly in developing nations where public health resources are minimal. The ability for these organisms to rapidly adapt to altered environments, particularly those where anti-bacterial drugs or vaccines exist, is a major cause of the ongoing problem. Here we present an overview of examples that demonstrate the impact that insertion sequence-induced gene activation can exert in human bacterial pathogens. Due to the ease in identifying drug-resistant bacterial mutants, the majority of our examples relate to drug resistance. However, several examples also exist of other virulence characteristics being enhanced and we have also provided documentation of these. We hope to demonstrate that along with the negative gene expression effects typically attributed to insertion sequence integration mutations, many of these elements generate diverse phenotypic effects by upregulating genes. The influence that this phenomenon has on human bacterial pathogenesis and, in particular, drug resistance is emphasised.

Bacterial Promoters

The main element involved in the regulation of gene transcription is the promoter. The ribonucleic acid (RNA) polymerase enzyme interacts with a DNA sequence, known as the promoter, upstream of the transcription start point (TSP) of a gene to allow initiation of transcription and thereby expression of the gene. Promoters usually occur upstream of, and outside the coding region of a gene. They can, however, also occur either within the coding region of an adjacent upstream gene, or within the coding sequence downstream of the TSP.

In bacteria, DNA is transcribed to mRNA by an RNA polymerase holoenzyme consisting of four subunits, structured $\alpha_2\beta\beta'\sigma$. The σ subunit (known as the sigma factor) recognises and binds to specific DNA promoter sequences, while the $\alpha_2\beta\beta'$ core complex carries out RNA polymerization. All bacterial species posses an essential, principle σ^{70}-type sigma factor which recognizes the promoters of constitutively expressed genes. Most bacteria have, in addition to the principle σ^{70}-type sigma factor, other alternate sigma factors which control conditional gene expression by coordinating the transcription of groups of genes or regulons. Their expression is restricted to specific growth conditions. All sigma factors initiate transcription by binding to the same core RNA polymerase but the different families recognize different promoter consensus sequences. In addition to alternative sigma factors, other proteins called transcription regulators, or transcription factors, can bind to certain other DNA sequences in and around the promoter thereby up- or down- regulating transcription and altering the expression of the gene.

The classical prokaryotic promoter structure recognized by σ^{70} sigma factors has been defined based on studies of numerous *Escherichia coli* promoters and similar sequences have since been identified in other bacteria (Hawley and McClure, 1983). The transcription start point (TSP) has been assigned position +1 and is usually a purine base; adenine or guanine. The classical promoter consists of a highly conserved hexamer centered about 10 bp upstream of the TSP, called the -10 sequence or Pribnow box. Another conserved hexamer is found 35 bp upstream of the TSP and is called the -35 sequence. The *E. coli* -35 and -10 hexamers have the consensus sequence TTGACA and TATAAT respectively (Figure 3). Promoter recognition studies suggest that two domains of the σ^{70}-type sigma factor interact with the -10 promoter sequence while an additional σ^{70} domain interacts with the -35 sequence (Gruber and Gross, 2003). Changes in the sequence of these conserved regions alters the strength of association between the RNA polymerase and the DNA thereby altering expression of the gene. The -10 and -35 hexamers are separated by a spacer, or interhexameric region, with the optimal length 17±1bp. Although the sequence between the -10 and -35 hexamers appears to play no major role in transcription of constitutive promoters, it may contain a transcription regulator binding site, thereby affecting the expression of inducible genes. Translation usually begins at an ATG 'start' codon, translating a methionine amino acid. The Shine-Dalgarno sequence, also known as the ribosome-binding site (RBS), is a purine-rich sequence of 3 to 10 nucleotides, situated 3' of the TSP but 4 to 7 nucleotides upstream of the start codon (Figure 3). Base pairing of the RBS in the mRNA and the 16S rRNA of the ribosome allows the ribosome to select the correct start codon and translate the mRNA to its corresponding polypeptide. The DNA sequence of the promoter is thus essential for optimal gene expression and various changes to the sequence can either induce or inhibit expression both constitutively and under specific growth conditions.

Mechanisms of Insertion Sequence-Mediated Gene Activation

The transposition of bacterial insertion sequences may increase gene expression by means of five general methods that may be divided into two groups (Figure 4). In this classification scheme, group one comprises mechanisms that directly activate gene expression from an insertion sequence-derived promoter (direct activation), while the second group relies on the insertional mutagenesis or displacement of an inhibitory transcriptional element by the insertion sequence in order to produce increased gene transcription (indirect inactivation).

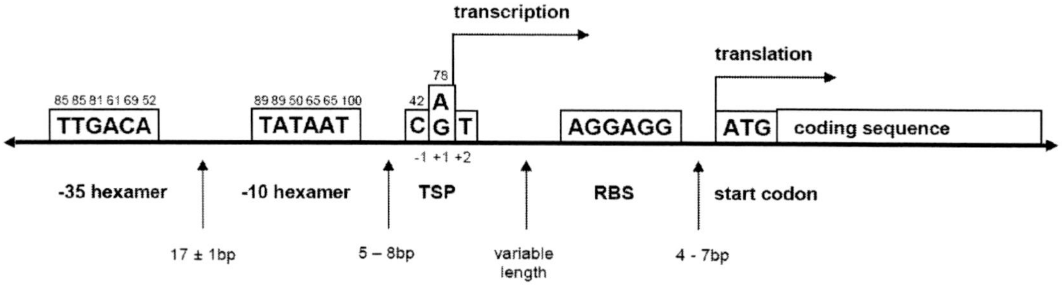

Figure 3. Consensus bacterial promoter sequence. The numbers above the sequence represent the likelihood that the base occurs at that specific position. TSP: Transcription start point; RBS: Ribosomal binding site.

Group 1. Direct Activation

Transcriptional activation of a gene may result from an insertion sequence possessing fully functional promoter activity within its 3' end that is able to mediate outward directed transcription (Figure 4a). Transcription may be initiated when such an element integrates upstream of, and in the correct orientation to, operons or individual genes, and these insertion sequences may be considered to be fully functional mobile promoters. Transposable elements may also induce gene transcription through the formation of hybrid promoters (Figure 4b). These promoters comprise sequences from both the insertion sequence and the target DNA and can occur when an insertion sequence integrates between the -10 and -35 conserved hexamers of a pre-existing promoter. If the 3'-terminal insertion sequence DNA carries sequences partially homologous to the consensus -35 hexamer and if this sequence is at an appropriate distance (approximately 17 bp) from the -10 hexamer, a new promoter, often possessing enhanced activity, may form. This appears to be the most commonly reported mechanism of insertion sequence-induced gene activation. In some cases, a hybrid promoter may be created upstream of a pre-existing promoter, resulting in a gene that is influenced by multiple promoters. An example of this is described by Olliver and colleagues (Olliver et al., 2005) and is shown diagrammatically in Figure 5c. Some insertion sequences, such as IS (insertion sequence) *3* and IS*257* have been shown to activate genes by both direct activation mechanisms (Aronson et al., 1989) (Simpson et al., 2000). Figure 6 shows detail of the promoter organisation in such an example. An added layer of complexity may arise if the activated gene is itself a transcription (or otherwise activating) factor for another gene. A mechanism of activation of the *bgl* operon, which allows β-glucoside utilisation by *E. coli*, seems likely to involve the integration of IS*10R* upstream of the *bglJ4* gene. This gene shows homology to transcriptional activators and its probable activation by IS*10R* correlates with increased transcription of *bgl* proteins (Giel et al., 1996).

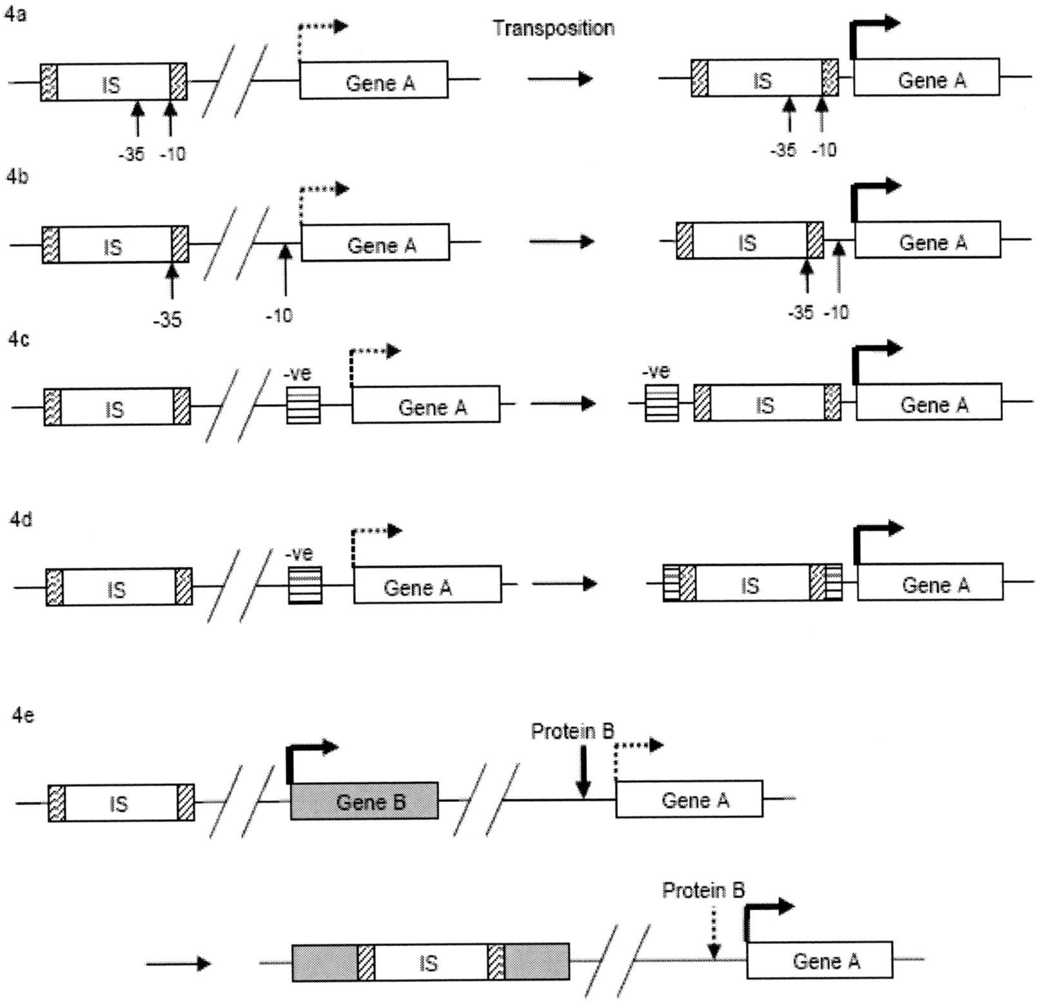

Figure 4. Mechanisms of insertion sequence-mediated transcriptional activation. Simplified schematic representation of the mechanisms leading to the transcriptional activation of a gene (Gene A) by an insertion sequence (IS). Diagonally filled boxes represent the terminal IRs of the IS. 4a: The insertion sequence contains a fully functional outward directed promoter, possessing both -35 and -10 conserved hexamers, within the 3' end of the element that is able to stimulate transcription when integrated upstream of a gene. 4b: The insertion sequence contains a -35 conserved hexamer at its 3' end which is able to combine with the -5 hexamer of a pre-existing promoter to form a hybrid promoter with enhanced transcription ability. 4c: The insertion sequence integrates between a gene and a negative cis-acting regulatory element located upstream. Displacement of the element to a more distal position results in enhanced transcription. 4d: The insertion sequence integrates within the upstream negative cis-acting regulatory element of a gene disrupting its function and resulting in enhanced gene transcription. 4e: Integration of an insertion sequence into a negative trans-acting regulatory factor (Gene B, shaded) disrupts the gene and inhibits production of the regulatory protein (protein B) resulting in increased expression of Gene A.

Group 2. Indirect Activation

The second group may be divided into three closely related mechanisms, neither of which rely on the inherent promoter or hybrid-promoter activity of an insertion sequence. Firstly, genes may be activated by insertion sequence-mediated displacement of negative *cis*-acting regulators in a pre-existing promoter (Figure 4c). In this case the negative regulator is relocated from the immediate vicinity of the promoter to a position upstream of the newly integrated insertion sequence where its influence is greatly reduced. Secondly, insertion sequence integration within a negative *cis*-acting regulator in a pre-existing promoter can also disrupt its function and result in increased gene transcription (Figure 4d). Finally, insertional mutagenesis into a gene coding for a negative *trans*-acting regulatory protein can result in increased expression of a previously negatively regulated gene (Figure 4e).

Activation of Genes Involved in Substrate Catabolism

The ability to metabolise or utilise alternative substrates as energy sources may provide a selective advantage for bacteria. To our knowledge, no bacterial pathogen has been shown to use such functions as a mechanism to increase its infectious or virulence related capabilities. However, the use of alternate metabolites via insertion sequence-induced gene expression has been documented in some organisms known to be potential human pathogens. For example, the ability of *E. coli* to metabolise threonine (known as the Tuc$^+$ phenotype) has been shown to be insertion sequence-mediated. Wild-type *E. coli* is unable to metabolise threonine and its metabolism is dependent on activation of the *tdh* operon. In a study of 8 independently generated Tuc$^+$ strains the increased expression of the *tdh* operon was found to be caused by the insertion of IS*3* into the promoter region of the operon. This formed an unregulated hybrid promoter that produced a 100-fold increase in expression relative to wild-type (Aronson et al., 1989). The *bgl* operon of *E. coli*, which allows for the metabolism of the β-glucoside sugars arbutin and salicin, is also upregulated by insertion sequences. This operon is generally silenced by flanking negative regulators. Disruptions to these transcriptional inhibitory factors may involve insertional mutagenesis by several distinct insertion sequences (Schnetz and Rak, 1988). As described above, the *bgl* operon may also be activated by a more complex insertion sequence promoter-mediated mechanism (Figure 4e) (Giel et al., 1996). Interestingly, insertion sequence activation of the *bgl* operon has been shown to be an example of adaptive evolution, i.e. the mutations do not display random specificity but appear to be targeted, beneficial mutations, that only occur under specific selective conditions (Hall, 1998). The *E. coli cel* operon, which is also involved in β-glucoside metabolism, appears to operate in a similar fashion (Parker and Hall, 1990), as does the adenine deaminase (*ade*) gene that is involved in the conversion of adenine to guanine (Petersen et al., 2002).

An example from another organism comes from *Burkholderia cepacia*, which displays pathogenicity in human cystic fibrosis. *B. cepacia* is an extremely metabolically versatile bacterium and a strain has been produced that is capable of utilising the herbicide 2,4,5-

trichlorophenoxyacetic acid (2,4,5-T) as its sole carbon source (the only known organism to metabolise 2,4,5-T in pure culture). This ability was demonstrated to derive from the integration of IS*1490* 110 bp upstream of the *tftAB* gene which encodes a dioxygenase required for the initial step in 2,4,5-T metabolism. A hybrid promoter is produced by this integration and constitutive gene transcription results (Hubner and Hendrickson, 1997). Nine other insertion sequences were found to reside within the *B. cepacia* genome and several of these are linked to gene clusters involved with other metabolic processes. The authors propose that these transposable elements might contribute to the extreme phenotypic diversity and metabolic versatility of this organism (Hubner and Hendrickson, 1997).

Activation of Genes Involved in General Virulence Properties

Bacterial virulence, defined as the ability to cause disease, is essential for a pathogenic organisms reproductive success and is therefore under strong selection pressure. Aspects of virulence are numerous and include transmission ability, growth or colonisation rate, toxin production, host immune inhibition and/or evasion, host cell invasion ability, and ability to resist anti-microbial drugs or vaccines. Since these characteristics are under genetic control the potential for insertion sequence gene activation-related virulence enhancement exists.

The potential importance of insertion sequences in the evolution of bacterial virulence has been implied in several studies (Parkhill et al., 2003; Chain et al., 2004; Kenna et al., 2006; Iwobi et al., 2002). In the Bordetella genus, for example, the ancestral species *B. bronchiseptica*, which rarely infects humans, has the largest genome (5.34 Mb) but possesses no insertion sequences. This contrasts with the more recently evolved *B. pertusis*, the agent of whooping cough, which, despite having a much smaller genome (4.1 Mb) possesses over 260 insertion sequences, including 238 copies of IS*481* (Parkhill et al., 2003). The precise genetic consequences of this spectacular example of insertion sequence expansion and its relationship to genome reduction and phenotypic alteration in the organism is currently poorly understood and apart from drug resistance (see below), few specific examples of insertion sequence-mediated transcriptional activation resulting in enhanced virulence are present in the literature. This undoubtedly relates to difficulties in the detection, isolation and genetic characterisation of drug susceptible hypervirulent bacterial strains (compared to the relative ease of bacterial selection for drug resistance) rather than any intrinsic bias for insertion sequence influence on drug resistance determining genes. It is likely therefore that many additional examples to those described below await documentation. Below we summarise several specific examples where insertion sequence-mediated transcriptional activation has increased bacterial virulence properties.

An example of gene activation caused by insertion sequence-mediated displacement of a negatively regulation *cis* transcription factor (Figure 4c) is provided by a mechanism of hyperleukotoxin production in *Actinobacillus actinomycetemcomitans* (Mitchell et al., 2003). *A. actinomycetemcomitans* is an oral commensal associated with infections of the oral cavity. Its virulence is partly attributed to the expression of a leukotoxin that destroys polymorphonuclear leukocytes and monocytes and severe forms of perionatal disease have been associated with strains exhibiting enhanced leukotoxin production (Guthmiller et al., 2001). The leukotoxin is expressed from an operon containing the four genes *ltxC, ltxA, ltxB*, and *ltxD*, and an open reading frame (*orfA*) that encodes a polypeptide of unknown function that is cotranscribed with *ltxCABD* is located upstream of the operon. Considerable variation in leukotoxin expression is found between *A. actinomycetemcomitans* strains, with most expressing relatively low levels. He and colleagues (He et al., 1999) described a hyperleukotoxic strain that contained an IS*1301* insertion upstream from *orfA* along with a spontaneous IS*1301*⁻ mutant of this strain that had lost the hyperleukotoxic phenotype. Genetic analysis of the *ltx* operon promoter region and IS*1301* insertion site found that the integration of the insertion sequence did indeed upregulate the operon but that, suprisingly, it did not produce any outward directed promoter activity. Instead, it was found that IS*1301* had integrated between a positive *cis*-acting transcriptional regulator located downstream from the insertion site, and a negative *cis*-acting transcriptional regulator located upstream from the insertion site. It is predicted that the relocation of the negative *cis*-acting regulator to over 900 bp upstream from the basal *ltx* operon promoter elements results in the increased leukotoxin production (Mitchell et al., 2003). IS*1310* integrates at a specific DNA sequence and it may be speculated that the presence of this sequence in such a crucial region for leukotoxicity production may be due to positive selection pressure. Indeed, Demuth and colleagues suggest that IS*1310* functions as a reversible genetic switch that, via its integration and excision from this region, can modifiy leukotoxin production depending on the needs of the organism (Mitchell et al., 2003).

Another example of insertion sequence-mediated gene activation resulting in an increased virulence phenotype involves the IS*6110* element found within members of the *Mycobacterium tuberculosis* complex (MTBC). *M. tuberculosis* is the causative agent of tuberculosis, arguably the leading cause of infectious disease mortality in the world. IS*6110* is an IS*3*-related transposable element that, due to its high inter-strain polymorphism (in both a numerical and positional sense), is widely used as a molecular marker for epidemiological studies. Its ability to act as a fully functional mobile promoter and thus influence gene expression and phenotype has only become apparent relatively recently however (Safi et al., 2004) (Soto et al., 2004). The ability of IS*6110* to increase the virulence of *M. tuberculosis* has been demonstrated by Soto and colleagues who have described the effects of its insertion upstream of the *phoP* gene (Soto et al., 2004). PhoP is a transcriptional regulator and knockout mutants result in impaired *M. tuberculosis* growth in mouse macrophages and attenuation in an *in vivo* mouse infection model. It is thus considered to be essential for *M. tuberculosis* virulence (Perez et al., 2001). Genetic examination of an outbreak strain of *M. bovis* (an MTBC member extremely closely related to *M. tuberculosis*) demonstrated the presence of an IS*6110* element 75 bp upstream from the ATG start codon of *phoP*. This insertion was associated with with an approximate 10-fold increase in phoP transcription compared to wild-type when the *phoP* gene and promoter region were cloned into a mycobacterial replicative plasmid and expressed in *M. smegmatis*. The authors conclude that the IS*6110*-associated transcriptional increase of *phoP* is likely to be strongly associated with high virulence (Soto et al., 2004).

The alteration of antigenic composition can also provide survival advantages for infectious bacteria. Evidence suggesting that insertion sequence-mediated gene activation may result in antigenic alteration has been described in the widespread human pathogen *Leptospira interrogans* (Zuerner and Trueba, 2005). Here, the genomes of two spontaneous mutants generated by immune selection were found to have undergone replicative IS*1501* transposition events resulting in additional insertion sequence copies in the mutants. Analysis of transcripts at altered loci demonstrated that IS*1501* could function as an outward directed promoter. Although a correlation between altered antigenicity and insertion sequence transposition and promoter activity was not attempted, these results hint at an additional virulence-related mechanism able to be modulated by insertion sequence generated gene activation.

An additional interesting phenotypic effect of insertion sequence-mediated gene upregulation involves the *E. coli* flagellum and its role in bacterial motility. Flagellae provide bacteria with an increased ability to control their environment by allowing movement away from poor conditions towards those more favorable. In *E. coli*, the *flhD* operon is the master operon of the flagellar regulon and bacterial motility is highly correlated with the presence of insertion sequences in the *flhD* promoter and subsequent increased levels of *flhD* transcript (Barker et al., 2004). Both IS*1* and IS*5* *flhD* promoter insertions have been shown to increase flagellar motility and this appears to result from a decrease in transcriptional repression since the formation of hybrid promoters were not responsible.

Activation of Genes with Antibiotic Inactivation Properties

Bacterial resistance to antimicrobial drugs is widespread and often occurs alarmingly soon after the introduction of the drug into a clinical setting. The increasing, often indiscriminate, use of antibiotics over the last two decades has fuelled this worrying phenomenon. A major source of the problem lies in the ability of microbes to evolve extremely rapidly. This is due to both their rapid reproductive rate and the variety of genetic mechanisms used to generate altered phenotypes. These include both inter-chromosomal mutations and the acquisition of extra-chromosomal drug resistance determining genes, often in the form of plasmid-bourne composite transposons, from other organisms either of the same or different species (Fig 2). These transposons contain gene cassettes that are often comprised of genes involved in drug resistance and they are capable of undergoing HGT. Insertion sequences may be involved in both of these processes. They have been found to be essential components of bacterial composite transposons, generally flanking the transposon and mediating bi-directional mobilisation between chromosomal and extra-chromosomal DNA (Fig 2). Numerous examples of transposon-associated insertion sequence elements positioned upstream of antibiotic resistant determining genes exist but their impact on downstream gene transcription has generally not been investigated and, although exceptions are described below, it is assumed that the insertion sequences usually function simply as mobilisers. It is largely through these remarkable elements that multi-drug resistance has rapidly spread between many bacterial species (Galimand et al., 2005). The ability for a simple insertion sequence transposition event within a chromosome to result in a drug resistant phenotype (via both insertional mutagenesis or gene activating mechanisms) has also been documented many times. Below, we summarise, according to drug family, some of the instances where drug resistance has been caused by either inter-chromosomal or extra-chromosomal insertion sequence-induced gene upregulation. The upregulation of efflux pumps, which may simultaneously provide resistance to a wide variety of antibiotic families, is discussed separately.

Resistance to β-Lactams

The β-lactam ring structure is a defining characteristic of the β-lactams, the most widely used group of antibiotics. These principally comprise the penicillin, carbapenem, cephalosporin, cephamycin and monobactam antibiotic families. These compounds function through their inhibition of bacterial cell wall synthesis and are particularly effective on gram positive bacteria. Resistance to β-lactams is often achieved by the expression of β-lactamases. This large family of enzymes all produce β-lactam antibiotic resistance by hydrolysing the β-lactam ring, thus rendering the compound inactive. β-lactamase upregulation due to insertion sequence-related promoter activity was first reported in *E. coli* in 1983 (Jaurin and Normark, 1983) and numerous examples now exist in the literature. One of the most thoroughly studied examples is carbapenem resistance in *Bacteroides fragilis*, an important anaerobic pathogen that is responsible for around 90% of anaerobic peritoneal infections. Around 3% of *B. fragilis* isolates carries a copy (or copies) of the metallo-β-lactamase carbapenem resistance gene, *cfiA*, which appears particularly susceptible to insertion sequence-mediated activation (Podglajen et al., 1995). *cfiA* has been shown to have the potential to be activated, with resultant β-lactam resistance, by around 10 distinct elements (Podglajen et al., 1994; Podglajen et al., 2001; Kato et al., 2003; Walsh et al., 2005). Most are members of the IS*4* family and appear to possess fully functional promoter sequences in their 3' IR region. Genetic analysis of *B. fragilis* strains has demonstrated a division into two robust monophyletic families which correspond to *cfiA* positive and *cfiA* negative groups. Interestingly, insertion sequences predominate in the smaller *cfiA* positive group suggesting their recent acquisition and subsequent coevolution (Ruimy et al., 1996).

Clinical isolates of the common opportunistic pathogen *Acinetobacter baumannii* exhibit a high rate of β-lactamase production and this appears to frequently involve insertion sequences. Several groups have documented increased resistance to β-lactams due to the integration of IS*1133*-like elements (designated IS*Aba1*) upstream of the bla_{OXA} family of carbapenemases (Turton et al., 2006; Hu et al., 2007; Zhou et al., 2007; Segal et al., 2007) and the cephalosporinase gene, *ampC* (Corvec et al., 2003; Segal et al., 2004; Heritier et al., 2006). The characterisation of the $bla_{OXA\text{-}23}$-associated IS*Aba1* promoter has recently been described (Segal et al., 2007). Here, IS*Aba1* was found to possess a fully functional promoter containing a -35 hexamer in addition to an extended -10 promoter sequence. Mammeri and colleagues have described the interspecies transfer of insertion sequence-mediated resistance to penicillins due to the integration of the *ampC* gene (possessing an IS*6*-related hybrid promoter) from *A. baumannii* into the chromosome of the low grade pathogen *Oligella urethralis* (Mammeri et al., 2003). The mechanism of this interspecies genetic transfer event remains unknown but may also be insertion sequence-mediated.

Interspecies transfer of antibiotic resistance genes may occur via the HGT of extrachromosomal elements. The integron-borne extended spectrum β-lactamase bla_{VEB-1} gene is found as a member of an extended gene cassette and its expression is driven by an IS*1999* or IS*2000* element (Aubert et al., 2003). This integron, In*50*, (Naas et al., 1999) and its resultant drug resistant phenotype has been observed in *Escherichia coli, Pseudomonas aeruginosa, Enterobacter cloacae, Enterobacter sacazakii,* and *Klebsiella pneumoniae* (Girlich et al., 2001; Girlich et al., 2002). In*50* may also be found chromosomally encoded, as is the usual case for *P. aeruginosa* (Girlich et al., 2002). IS*1999* is also capable of upregulating the plasmid-borne bla_{OXA-48} gene in *K. pneumoniae* (Poirel et al., 2004). In this case the resistant isolate contained 5 additional plasmid-borne β-lactamase genes and was found to be resistant to all 14 β-lactams tested. An isolate of *K. pneumoniae* has also been shown to employ IS*26* as a hybrid promoter to transcribe a plasmid-borne resistance operon containing two genes (Lee et al., 1990). One of these genes, *blaS2A*, encodes a β-lactamase that provides resistance to broad-spectrum cephalosporins. The other gene conferred resistance to aminoglycosides (see below). Besides In*50*-derived β-lactam resistance, *P. aeruginosa* has also been shown to undergo insertion sequence-derived insertional inactivation of a negative β-lactamase regulator with resultant increased β-lactamase expression and drug resistance (Bagge et al., 2002).

Finally, the IS*256* element usually found in *Staphlococcus aureus* has been shown to be capable of upregulating both the *llm* and *mecA* genes with resultant methicillin resistance (Maki and Murakami, 1997; Couto et al., 2003). The latter case involved IS*256*-mediated *mecA* overproduction in *S. sciuri*, the most common Staphylococcus species worldwide and an organism only distantly related to *S. aureus*. IS*256* is not commonly found in *S. sciuri* and the authors suggest that it was acquired from an exogenous source and was subsequently exploited for its ability to form a powerful hybrid promoter for *mecA* (Couto et al., 2003).

Resistance to Aminoglycosides

Aminoglycosides act by binding to the bacterial 30S or 50S ribosomal subunit, thereby inhibiting peptidyl-tRNA translocation and resulting in the inhibition of protein synthesis. Several genes have been documented that encode aminoglycoside modifying enzymes and these are generally located on extrachromosomal elements. Multidrug-resistant isolates of *Actinobacter baumannii* usually contain plasmid-borne integrons and transcription of the aminoglycoside gene *aac(3)-IIa*, present on these integrons, is suggested to be mediated by an IS*1133*-derived hybrid promoter (Segal et al., 2003). Chromosomal aminoglycoside modifying enzymes such as the *aac(6')-Ij* and *aac(6')-Ig* aminoglycosine acetyltransferase genes of *Actinobacter* sp. 13 have also been documented (Lambert et al., 1994). Rudant and colleagues have described an aminoglycoside susceptible strain of *Acinobacter* sp. 13 and the activation of its silent *aac(6')-Ig* gene by integration of IS*18* upstream of the gene (Rudant et al., 1998). The resultant hybrid promoter provided sufficient transcriptional activity to produce resistance to various aminoglycosides. The study of Lee and colleagues described above (Lee et al., 1990) of a plasmid-borne antibiotic resistance operon in *Klebsiella pneumoniae* has also demonstrated insertion sequence-mediated upregulation of an aminoglycoside resistance conferring enzyme. In this case an IS*26* insertion was found to upregulate the operon which includes the *aphA7* gene which confers resistance to neomycin and other structurally related aminoglycosides.

Resistance to Glycopeptides

Glycopeptides are a class of antibiotic drugs that function by inhibiting bacterial cell wall synthesis. They achieve this by binding to amino acids within the cell wall and inhibiting the addition of new units to the peptidoglycan polymer. They are principally effective against gram positive bacterium and, due to their toxicity, are generally used as antibiotics of last resort in patients that are β-lactam hypersensitive or against bacteria that demonstrate resistance to other drugs. Vancomycin resistance in *Enterococcus* species is mediated by the plasmid-borne Tn*1546*, Tn*1547*, or Tn*1549* transposons. The most frequently encountered transposon, Tn*1546*, codes for 9 polypeptides including the *vanRSHAXYZ* gene cluster and is highly diverse, containing various point mutations along with numerous differences in insertion sequence content and location (Willems et al., 1999). The insertion sequences of Tn*1546* likely play a central role in its evolution and can sometimes disrupt gene function. However, they have also demonstrated their ability to upregulate expression of the resistance conferring genes through hybrid promoter formation (Brown et al., 2001). Insertion sequences are also sometimes found in other intergenic regions on Tn*1546* but their potential promoter activities have not been investigated.

Resistance to Tetracyclines

Tetracyclines act by binding to the 16S section of the 30S ribosomal subunit and inhibiting the amino-acyl tRNA from binding to the ribosome. Their bacteriocidal properties thus relate to their prevention of mRNA translation. Tetracycline resistance genes are generally associated with insertion sequences that reside on plasmids and/or transposons. Many transposon families carrying a variety of resistance genes have been described and, as for glycopeptide resistance genes, these differ greatly in the type, number and position of their associated insertion sequences. Insertion sequence-related influence on the transcription of tetracycline resistance genes in these transposases has not been greatly explored, but the studies of Simpson and colleagues (Simpson et al., 2000) has shown that it does exist. In this example, the transcription of a *tetA(k)* tetracycline resistance gene that resides in a chromosomal plasmid cointegrate of a tetracycline resistant strain of *S. aureus* was investigated. The authors found that IS*257* formed a hybrid promoter with *tetA(k)* that produced a far stronger promoter than the wild type. Interestingly, apart from its ability to form hybrid promoters, IS*257* also appears to possess a complete internal outwardly-directed promoter (Fig 6). Furthermore, it also seems likely to be capable of upregulating the dihydrofolate reductase inhibitor trimethoprim in *Staphlococcii* (see below) (Leelaporn et al., 1994). Several additional potential IS*257*-derived promoters have been located upstream of genes encoding resistance to aminoglycosides, cadmium, and mupirocin (Simpson et al.,

2000). Since IS*257* also plays a major role in the incorporation of resistance determinants into plasmids and chromosomes it appears that this mobile element plays a particularly active role in generating diverse resistance characteristics in *Staphlococcii.*

Resistance to Nitroimidazoles

Nitroimidazoles are chemically reduced by anaerobic bacteria to an active form that disrupts DNA structure and prevents nucleic acid synthesis. Members of the *Bacteroides sensu stricto* genus often possess several types of nitroimidazole resistance genes (designated *nimA-G*) that are pedicted to be activated by insertion sequences located upstream (Haggoud et al., 1994; Trinh et al., 1995; Nagy et al., 2001; Soki et al., 2006). Multiple *nim* genes may be present in any single Bacteroides strain and these genes display differences in their preferred localisation (ie chromosome or plasmid) and associated mobile element. For example, *nimB* is generally chromosomal, whereas *nimE* is plasmid-borne, and IS*1168* associates with *nimA* and *nimB* while IS*1170* associates with *nimC* (Soki et al., 2006). These associations are not always strictly observed and it should also be noted that the promoter activity of these insertion sequences is yet to be experimentally confirmed.

Resistance to Trimethoprim

Trimethoprim is a dihydrofolate reductase inhibitor that prevents the synthesis of bacterial tetrahydrofolic acid. This is an essential precursor of thymidine and its loss results in an inability to produce this nucleotide for DNA synthesis. Many drug resistance-conferring dihydrofolate reductases have been characterised and these can be spread as gene cassettes in integrons on plasmid-bourne composite transposons. Although these genetic constructs are associated with many different insertion sequences, few studies have attempted to describe their relation to gene expression. IS*257* has, however, been shown to upregulate the dihydrofolate reductase gene *dfrA* in *S. aureus*, resulting in low-level Trimethoprim resistance (Leelaporn et al., 1994).

Resistance to Sulfonamides

In bacteria, sulfonamides function as competitive inhibitors of dihydropteroate synthetase. This enzyme catalyses a key step in folate synthesis and its loss prevents the synthesis of nucleic acids. IS*Aba-1* has been shown to contain a promoter that is capable of upregulating the *ampC* and bla_{OXA} β-lactamase genes in *A. baumannii* (see above). This insertion sequence has also been found located upstream of the sulphonamide resistance gene *sulII* in an *A. baumannii* strain that displayed sulphonamide resistance (Segal et al., 2005). The promoter activity of IS*Aba-1* has not been confirmed in this case and is only assumed.

Efflux Pump-Mediated Resistance

Efflux pumps perform the important physiological role of removing various harmful compounds from the cell. Therefore, by their very design, antibiotics are targets for efflux and it is not suprising that antibiotic resistant bacteria often display increased efflux ability. This specific form of resistance has a particularly important clinical relevance because efflux pump overexpression often acts with low specificity, resulting in multidrug resistance. The mechanisms involved in efflux pump-mediated drug resistance may be divided into two distinct categories: i) the expression of an efflux pump gene is increased due to either an alteration in its promoter region or to the promoter region of a gene that, in turn, affects efflux pump gene transcription and ii) a mutation in the coding region of the gene results in the increased efflux ability of the corresponding protein. In the former scenario, the ability of insertion sequences to provide the necessary promoter for increased efflux pump gene transcription has been documented in several cases.

E. coli, for example, contains a large number of characterised efflux pumps, including *acrAB*, *emrAB*, *mdfA*, *acrD*, and *acrEF* that are known, or suspected, to be involved in drug resistance. Jellen-Ritter and colleagues, in a series of efflux gene knockout experiments, described the activation of the *acrEF* operon in an *acrAB* knockout that was multidrug-resistant. *AcrEF* is only weakly expressed under normal laboratory conditions and its upregulation was caused by the insertion of IS*2* into the promoter and it is this upregulation that was proposed to have compensated for the deletion of *acrAB* (Jellen-Ritter and Kern, 2001). In a similar fashion, *acrEF* has also been shown to be upregulated by an IS*1* or IS*2* insertion upstream of the operon in *acrB* knockouts that are resistant to hydrophobic solvents (Kobayashi et al., 2001). This further emphasises the broad substrate specificity of acrEF. The *acrEF* operon appears to be particularly susceptible to this form of activation with *Salmonella enterica* also being shown to undergo insertion sequence intergration in this region (Olliver et al., 2005). In this case IS*1* or IS*10* was found to be integrated into the *acrEF* promoter region, resulting in fluoroquinolone resistance. The organisation of new promoter sequences generated in this example are shown in Figure 5. RT-PCR experiments revealed an 8 to 10-fold increase in *acrF* expression following their integration (Olliver et al., 2005).

Insertional inactivation of efflux repressor genes by insertion sequences has also been demonstrated (Boutoille et al., 2004; Webber and Piddock, 2001). In the case described by Boutoille and colleagues, IS*21* insertional inactivation of a gene involved in the repression of the *mexAB-OprM* operon efflux system in a clinical strain of *Pseudomonas aeruginosa* resulted in increased *mexA* transcription and subsequent resistance to ticarcillin and aztreonam (Boutoille et al., 2004).

Summary

Here we have attempted to provide a brief overview of instances where the transcriptional activation of genes in human bacterial pathogens has been mediated by the actions of an insertion sequence.

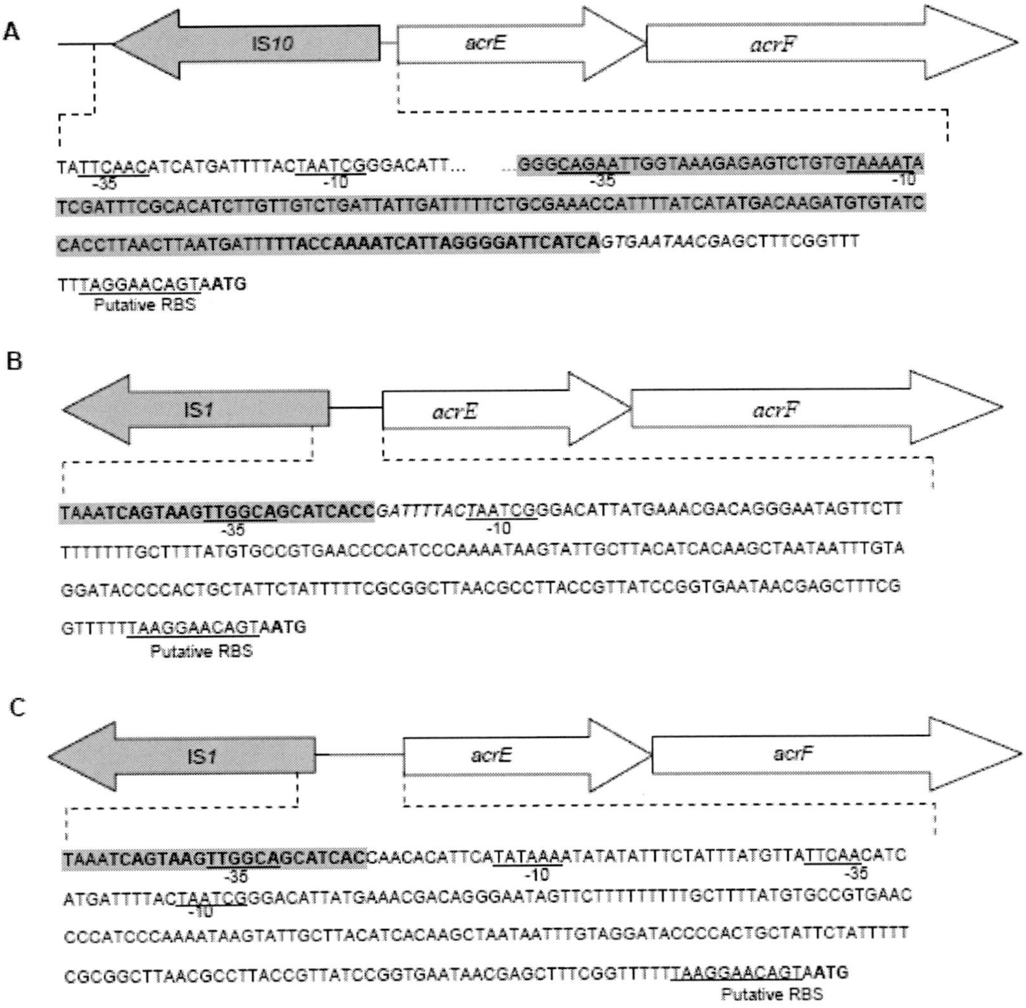

Figure 5. Schematic representation of new promoter constructs formed by the integration of IS*10* and IS*1* upstream of the *acrEF* operon. A: IS*10* (shaded sequence), containing an internal promoter, integrates between the pre-existing promoter sequence and the *acrEF* operon. -35 and -10 hexamers of the displaced pre-existing and IS-contained promoters are underlined. IR region of IS*10* and the *acrE* ATG start codon are shown in bold. 10 bp duplication of target sequence formed during integration shown in italyx. B: IS*1* integration between the *acrEF* -35 and -10 promoter regions forms a hybrid promoter containing the original -10 sequence and a -35 sequence derived from the IS*1* IR region. IS*1* shown as shaded sequence. IR region of IS*1* and the *acrE* ATG start codon are shown in bold. Nine bp duplication of target sequence formed during integration shown in italyx. Putative RBS underlined. C: IS*1* integration upstream of the acrEF operon promoter region creates a new hybrid promoter while leaving the pre-existing promoter intact. Shaded region shows IS*1* sequence. Hybrid promoter and original promoter -35 and -10 hexamers are underlined. IR region of IS*1* and the *acrE* ATG start codon are shown in bold. Putative RBS underlined. Figure adapted from Olliver et al (Olliver et al., 2005).

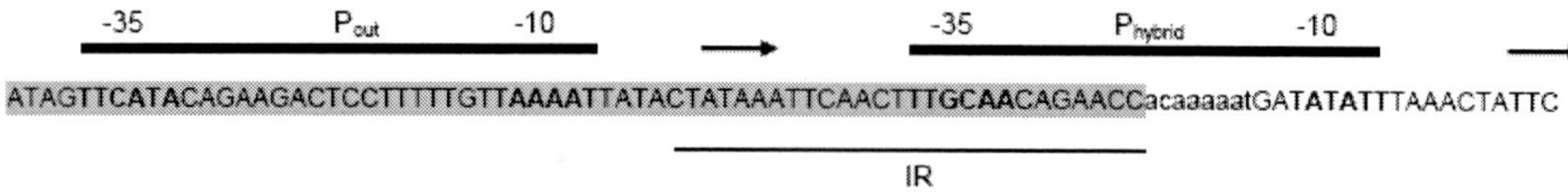

Figure 6. Promoter sequences of IS*257* located upstream of *tetA(K)* in *Staphylococcus aureus*. IS*257* contains sequences that can serve as a fully functional mobile promoter as well as a sequence that can function as the -35 hexamer in a hybrid promoter. The diagram above shows the genetic organisation of promoters following IS*257* integration upstream of the *tetA(K)* tetracycline resistance-determining gene in *S. aureus*. Two promoters are capable of upregulating transcription of *tetA(K)* in this example. The promoter Pout is contained entirely within the IS*257* and produces outward directed transcription. A hybrid promoter, Phybrid, is also formed by the fusion of IS*257* sequence containing a -35 hexamer within the 27 bp IR region, with a -10 hexamer located upstream of the integration site. This multiple promoter construct is estimated to increase total promoter strength by a factor of 60 compared to the wild-type promoter, mainly due to a more optimal -35 sequence in the hybrid promoter, and results in a high level of tetracycline resistance (Simpson et al., 2000). Shaded area represents IS*257* sequence. Bold sequence shows the -35 and -10 conserved hexamers of each promoter. Arrows represent the transcriptional start sites for the promoters. Lower case lettering indicates the 8 bp target DNA duplication sequence. Figure adapted from Simpson et al. (Simpson et al., 2000).

We have demonstrated that the phenotypic effects of this phenomenon are varied and may often aid the organism in its virulence characteristics, particularly regarding increased resistance to bacteriocidal drugs. Of note is the fact that many of these phenotypic alterations can be transferred to members of the same or different species and that this phenomenon is also largely dependent on the action of insertion sequences. When these factors are considered it is apparent that insertion sequences represent a major driving force in the generation of bacterial genomic diversity and plasticity, and can keep genomes in a constant state of flux. The evolutionary advantages that they provide to their hosts appear to overcome any disadvantages that might be linked to their parasitic tendencies, at least in the short term. This theory is strengthened by examples we have described that, for example, appear to demonstrate that fully functional insertion sequence promoters have been actively selected for, as have DNA promoter sequences that allow for the insertional inactivation of negative regulatory factors. Thus, even though some studies suggest that insertion sequences are evolutionary detrimental in the long term (Wagner, 2006), their positive phenotypic effects are very relevant in our time frame.

References

Aronson,B.D., Levinthal,M., and Somerville,R.L. (1989). Activation of a cryptic pathway for threonine metabolism via specific IS3-mediated alteration of promoter structure in Escherichia coli. *J. Bacteriol. 171*, 5503-5511.

Aubert,D., Naas,T., and Nordmann,P. (2003). IS1999 increases expression of the extended-spectrum beta-lactamase VEB-1 in Pseudomonas aeruginosa. *J. Bacteriol. 185*, 5314-5319.

Bagge,N., Ciofu,O., Hentzer,M., Campbell,J.I., Givskov,M., and Hoiby,N. (2002). Constitutive high expression of chromosomal beta-lactamase in Pseudomonas aeruginosa caused by a new insertion sequence (IS1669) located in ampD. *Antimicrob. Agents Chemother. 46*, 3406-3411.

Barker,C.S., Pruss,B.M., and Matsumura,P. (2004). Increased motility of Escherichia coli by insertion sequence element integration into the regulatory region of the flhD operon. *J. Bacteriol. 186*, 7529-7537.

Boutoille,D., Corvec,S., Caroff,N., Giraudeau,C., Espaze,E., Caillon,J., Plesiat,P., and Reynaud,A. (2004). Detection of an IS21 insertion sequence in the mexR gene of Pseudomonas aeruginosa increasing beta-lactam resistance. *FEMS Microbiol. Lett. 230*, 143-146.

Brown,A.R., Townsley,A.C., and Amyes,S.G. (2001). Diversity of Tn1546 elements in clinical isolates of glycopeptide-resistant enterococci from Scottish hospitals. *Antimicrob. Agents Chemother. 45*, 1309-1311.

Chain,P.S., Carniel,E., Larimer,F.W., Lamerdin,J., Stoutland,P.O., Regala,W.M., Georgescu,A.M., Vergez,L.M., Land,M.L., Motin,V.L., Brubaker,R.R., Fowler,J., Hinnebusch,J., Marceau,M., Medigue,C., Simonet,M., Chenal-Francisque,V., Souza,B., Dacheux,D., Elliott,J.M., Derbise,A., Hauser,L.J., and Garcia,E. (2004). Insights into the evolution of Yersinia pestis through whole-genome comparison with Yersinia pseudotuberculosis. *Proc. Natl. Acad. Sci. U. S. A 101*, 13826-13831.

Corvec,S., Caroff,N., Espaze,E., Giraudeau,C., Drugeon,H., and Reynaud,A. (2003). AmpC cephalosporinase hyperproduction in Acinetobacter baumannii clinical strains. *J. Antimicrob. Chemother. 52*, 629-635.

Couto,I., Wu,S.W., Tomasz,A., and de,L.H. (2003). Development of methicillin resistance in clinical isolates of Staphylococcus sciuri by transcriptional activation of the mecA homologue native to s. *J. Bacteriol. 185*, 645-653.

Doolittle,W.F. and Sapienza,C. (1980). Selfish genes, the phenotype paradigm and genome evolution. *Nature 284*, 601-603.

Frost,L.S., Leplae,R., Summers,A.O., and Toussaint,A. (2005). Mobile genetic elements: the agents of open source evolution. *Nat. Rev. Microbiol. 3*, 722-732.

Galimand,M., Sabtcheva,S., Courvalin,P., and Lambert,T. (2005). Worldwide disseminated armA aminoglycoside resistance methylase gene is borne by composite transposon Tn1548. *Antimicrob. Agents Chemother. 49*, 2949-2953.

Giel,M., Desnoyer,M., and Lopilato,J. (1996). A mutation in a new gene, bglJ, activates the bgl operon in Escherichia coli K-12. *Genetics 143*, 627-635.

Girlich,D., Naas,T., Leelaporn,A., Poirel,L., Fennewald,M., and Nordmann,P. (2002). Nosocomial spread of the integron-located veb-1-like cassette encoding an extended-pectrum beta-lactamase in Pseudomonas aeruginosa in Thailand. *Clin. Infect. Dis. 34*, 603-611.

Girlich,D., Poirel,L., Leelaporn,A., Karim,A., Tribuddharat,C., Fennewald,M., and Nordmann,P. (2001). Molecular epidemiology of the integron-located VEB-1 extended-spectrum beta-lactamase in nosocomial enterobacterial isolates in Bangkok, Thailand. *J. Clin. Microbiol. 39*, 175-182.

Gruber,T.M. and Gross,C.A. (2003). Multiple sigma subunits and the partitioning of bacterial transcription space. *Annu. Rev. Microbiol. 57*, 441-466.

Guthmiller,J.M., Lally,E.T., and Korostoff,J. (2001). Beyond the specific plaque hypothesis: are highly leukotoxic strains of Actinobacillus actinomycetemcomitans a paradigm for periodontal pathogenesis? *Crit Rev. Oral Biol. Med. 12*, 116-124.

Haggoud,A., Reysset,G., Azeddoug,H., and Sebald,M. (1994). Nucleotide sequence analysis of two 5-nitroimidazole resistance determinants from Bacteroides strains and of a new insertion sequence upstream of the two genes. *Antimicrob. Agents Chemother. 38*, 1047-1051.

Hall,B.G. (1998). Activation of the bgl operon by adaptive mutation. Mol. Biol. Evol. *15*, 1-5.

Hawley,D.K. and McClure,W.R. (1983). Compilation and analysis of Escherichia coli promoter DNA sequences. *Nucleic Acids Res. 11*, 2237-2255.

He,T., Nishihara,T., Demuth,D.R., and Ishikawa,I. (1999). A novel insertion sequence increases the expression of leukotoxicity in Actinobacillus actinomycetemcomitans clinical isolates. *J. Periodontol. 70*, 1261-1268.

Heritier,C., Poirel,L., and Nordmann,P. (2006). Cephalosporinase over-expression resulting from insertion of ISAba1 in Acinetobacter baumannii. *Clin. Microbiol. Infect. 12*, 123-130.

Hu,W.S., Yao,S.M., Fung,C.P., Hsieh,Y.P., Liu,C.P., and Lin,J.F. (2007). An OXA-66/OXA-51-Like Carbapenemase and Possibly an Efflux Pump Are Associated with Resistance to Imipenem in Acinetobacter baumannii. *Antimicrob. Agents Chemother. 51*, 3844-3852.

Hubner,A. and Hendrickson,W. (1997). A fusion promoter created by a new insertion sequence, IS1490, activates transcription of 2,4,5-trichlorophenoxyacetic acid catabolic genes in Burkholderia cepacia AC1100. *J. Bacteriol. 179*, 2717-2723.

Iwobi,A., Rakin,A., Garcia,E., and Heesemann,J. (2002). Representational difference analysis uncovers a novel IS10-like insertion element unique to pathogenic strains of Yersinia enterocolitica. *FEMS Microbiol. Lett. 210*, 251-255.

Jaurin,B. and Normark,S. (1983). Insertion of IS2 creates a novel ampC promoter in Escherichia coli. *Cell 32*, 809-816.

Jellen-Ritter,A.S. and Kern,W.V. (2001). Enhanced expression of the multidrug efflux pumps AcrAB and AcrEF associated with insertion element transposition in Escherichia coli mutants Selected with a fluoroquinolone. *Antimicrob. Agents Chemother. 45*, 1467-1472.

Kato,N., Yamazoe,K., Han,C.G., and Ohtsubo,E. (2003). New insertion sequence elements in the upstream region of cfiA in imipenem-resistant Bacteroides fragilis strains. *Antimicrob. Agents Chemother. 47*, 979-985.

Kenna,D.T., Yesilkaya,H., Forbes,K.J., Barcus,V.A., Vandamme,P., and Govan,J.R. (2006). Distribution and genomic location of active insertion sequences in the Burkholderia cepacia complex. *J. Med. Microbiol. 55*, 1-10.

Kobayashi,K., Tsukagoshi,N., and Aono,R. (2001). Suppression of hypersensitivity of Escherichia coli acrB mutant to organic solvents by integrational activation of the acrEF operon with the IS1 or IS2 element. *J. Bacteriol. 183*, 2646-2653.

Lambert,T., Gerbaud,G., and Courvalin,P. (1994). Characterization of the chromosomal aac(6')-Ij gene of Acinetobacter sp. 13 and the aac(6')-Ih plasmid gene of Acinetobacter baumannii. *Antimicrob. Agents Chemother. 38*, 1883-1889.

Lee,K.Y., Hopkins,J.D., and Syvanen,M. (1990). Direct involvement of IS26 in an antibiotic resistance operon. *J. Bacteriol. 172*, 3229-3236.

Leelaporn,A., Firth,N., Byrne,M.E., Roper,E., and Skurray,R.A. (1994). Possible role of insertion sequence IS257 in dissemination and expression of high- and low-level trimethoprim resistance in staphylococci. *Antimicrob. Agents Chemother. 38*, 2238-2244.

Mahillon,J. and Chandler,M. (1998). Insertion sequences. *Microbiol. Mol. Biol. Rev. 62*, 725-774.

Maki,H. and Murakami,K. (1997). Formation of potent hybrid promoters of the mutant llm gene by IS256 transposition in methicillin-resistant Staphylococcus aureus. *J. Bacteriol. 179*, 6944-6948.

Mammeri,H., Poirel,L., Mangeney,N., and Nordmann,P. (2003). Chromosomal integration of a cephalosporinase gene from Acinetobacter baumannii into Oligella urethralis as a source of acquired resistance to beta-lactams. Antimicrob. *Agents Chemother. 47*, 1536-1542.

Mitchell,C., Gao,L., and Demuth,D.R. (2003). Positive and negative cis-acting regulatory sequences control expression of leukotoxin in Actinobacillus actinomycetemcomitans 652. *Infect. Immun. 71*, 5640-5649.

Naas,T., Poirel,L., Karim,A., and Nordmann,P. (1999). Molecular characterization of In50, a class 1 integron encoding the gene for the extended-spectrum beta-lactamase VEB-1 in Pseudomonas aeruginosa. *FEMS Microbiol. Lett. 176*, 411-419.

Nagy,E., Soki,J., Urban,E., Szoke,I., Fodor,E., and Edwards,R. (2001). Occurrence of metronidazole and imipenem resistance among Bacteroides fragilis group clinical isolates in Hungary. *Acta Biol. Hung. 52*, 271-280.

Olliver,A., Valle,M., Chaslus-Dancla,E., and Cloeckaert,A. (2005). Overexpression of the multidrug efflux operon acrEF by insertional activation with IS1 or IS10 elements in Salmonella enterica serovar typhimurium DT204 acrB mutants selected with fluoroquinolones. *Antimicrob. Agents Chemother. 49*, 289-301.

Orgel,L.E. and Crick,F.H. (1980). Selfish DNA: the ultimate parasite. Nature *284*, 604-607.

Parker,L.L. and Hall,B.G. (1990). Mechanisms of activation of the cryptic cel operon of Escherichia coli K12. *Genetics 124*, 473-482.

Parkhill,J., Sebaihia,M., Preston,A., Murphy,L.D., Thomson,N., Harris,D.E., Holden,M.T., Churcher,C.M., Bentley,S.D., Mungall,K.L., Cerdeno-Tarraga,A.M., Temple,L., James,K., Harris,B., Quail,M.A., Achtman,M., Atkin,R., Baker,S., Basham,D., Bason,N., Cherevach,I., Chillingworth,T., Collins,M., Cronin,A., Davis,P., Doggett,J., Feltwell,T., Goble,A., Hamlin,N., Hauser,H., Holroyd,S., Jagels,K., Leather,S., Moule,S., Norberczak,H., O'Neil,S., Ormond,D., Price,C., Rabbinowitsch,E., Rutter,S., Sanders,M., Saunders,D., Seeger,K., Sharp,S., Simmonds,M., Skelton,J., Squares,R., Squares,S., Stevens,K., Unwin,L., Whitehead,S., Barrell,B.G., and Maskell,D.J. (2003). Comparative analysis of the genome sequences of Bordetella pertussis, Bordetella parapertussis and Bordetella bronchiseptica. *Nat. Genet. 35*, 32-40.

Perez,E., Samper,S., Bordas,Y., Guilhot,C., Gicquel,B., and Martin,C. (2001). An essential role for phoP in Mycobacterium tuberculosis virulence. *Mol. Microbiol. 41*, 179-187.

Petersen,C., Moller,L.B., and Valentin-Hansen,P. (2002). The cryptic adenine deaminase gene of Escherichia coli. Silencing by the nucleoid-associated DNA-binding protein, H-NS, and activation by insertion elements. *J. Biol. Chem. 277*, 31373-31380.

Pilacinski,W., Mosharrafa,E., Edmundson,R., Zissler,J., Fiandt,M., and Szybalski,W. (1977). Insertion sequence IS2 associated with int-constitutive mutants of bacteriophage lambda. *Gene 2*, 61-74.

Podglajen,I., Breuil,J., Casin,I., and Collatz,E. (1995). Genotypic identification of two groups within the species Bacteroides fragilis by ribotyping and by analysis of PCR-generated fragment patterns and insertion sequence content. *J. Bacteriol. 177*, 5270-5275.

Podglajen,I., Breuil,J., and Collatz,E. (1994). Insertion of a novel DNA sequence, 1S1186, upstream of the silent carbapenemase gene cfiA, promotes expression of carbapenem resistance in clinical isolates of Bacteroides fragilis. *Mol. Microbiol. 12*, 105-114.

Podglajen,I., Breuil,J., Rohaut,A., Monsempes,C., and Collatz,E. (2001). Multiple mobile promoter regions for the rare carbapenem resistance gene of Bacteroides fragilis. *J. Bacteriol. 183*, 3531-3535.

Poirel,L., Heritier,C., Tolun,V., and Nordmann,P. (2004). Emergence of oxacillinase-mediated resistance to imipenem in Klebsiella pneumoniae. *Antimicrob. Agents Chemother. 48*, 15-22.

Rudant,E., Courvalin,P., and Lambert,T. (1998). Characterization of IS18, an element capable of activating the silent aac(6')-Ij gene of Acinetobacter sp. 13 strain BM2716 by transposition. *Antimicrob. Agents Chemother. 42*, 2759-2761.

Ruimy,R., Podglajen,I., Breuil,J., Christen,R., and Collatz,E. (1996). A recent fixation of cfiA genes in a monophyletic cluster of Bacteroides fragilis is correlated with the presence of multiple insertion elements. *J. Bacteriol. 178*, 1914-1918.

Saedler,H., Reif,H.J., Hu,S., and Davidson,N. (1974). IS2, a genetic element for turn-off and turn-on of gene activity in E. coli. *Mol. Gen. Genet. 132*, 265-289.

Safi,H., Barnes,P.F., Lakey,D.L., Shams,H., Samten,B., Vankayalapati,R., and Howard,S.T. (2004). IS6110 functions as a mobile, monocyte-activated promoter in Mycobacterium tuberculosis. *Mol. Microbiol. 52*, 999-1012.

Schnetz,K. and Rak,B. (1988). Regulation of the bgl operon of Escherichia coli by transcriptional antitermination. *EMBO J. 7*, 3271-3277.

Segal,H., Garny,S., and Elisha,B.G. (2005). Is IS(ABA-1) customized for Acinetobacter? *FEMS Microbiol. Lett. 243*, 425-429.

Segal,H., Jacobson,R.K., Garny,S., Bamford,C.M., and Elisha,B.G. (2007). Extended -10 promoter in ISAba-1 upstream of blaOXA-23 from Acinetobacter baumannii. *Antimicrob. Agents Chemother. 51*, 3040-3041.

Segal,H., Nelson,E.C., and Elisha,B.G. (2004). Genetic environment and transcription of ampC in an Acinetobacter baumannii clinical isolate. *Antimicrob. Agents Chemother. 48*, 612-614.

Segal,H., Thomas,R., and Gay,E.B. (2003). Characterization of class 1 integron resistance gene cassettes and the identification of a novel IS-like element in Acinetobacter baumannii. Plasmid *49*, 169-178.

Simpson,A.E., Skurray,R.A., and Firth,N. (2000). An IS257-derived hybrid promoter directs transcription of a tetA(K) tetracycline resistance gene in the Staphylococcus aureus chromosomal mec region. *J. Bacteriol. 182*, 3345-3352.

Soki,J., Gal,M., Brazier,J.S., Rotimi,V.O., Urban,E., Nagy,E., and Duerden,B.I. (2006). Molecular investigation of genetic elements contributing to metronidazole resistance in Bacteroides strains. *J. Antimicrob. Chemother. 57*, 212-220.

Soto,C.Y., Menendez,M.C., Perez,E., Samper,S., Gomez,A.B., Garcia,M.J., and Martin,C. (2004). IS6110 mediates increased transcription of the phoP virulence gene in a multidrug-resistant clinical isolate responsible for tuberculosis outbreaks. *J. Clin. Microbiol. 42*, 212-219.

Tavakoli,N.P. and Derbyshire,K.M. (2001). Tipping the balance between replicative and simple transposition. *EMBO J. 20*, 2923-2930.

Trinh,S., Haggoud,A., Reysset,G., and Sebald,M. (1995). Plasmids pIP419 and pIP421 from Bacteroides: 5-nitroimidazole resistance genes and their upstream insertion sequence elements. *Microbiology 141 (Pt 4)*, 927-935.

Turton,J.F., Ward,M.E., Woodford,N., Kaufmann,M.E., Pike,R., Livermore,D.M., and Pitt,T.L. (2006). The role of ISAba1 in expression of OXA carbapenemase genes in Acinetobacter baumannii. *FEMS Microbiol. Lett. 258*, 72-77.

Wagner,A. (2006). Periodic extinctions of transposable elements in bacterial lineages: evidence from intragenomic variation in multiple genomes. *Mol. Biol. Evol. 23*, 723-733.

Walsh,T.R., Onken,A., Haldorsen,B., Toleman,M.A., and Sundsfjord,A. (2005). Characterization of a carbapenemase-producing clinical isolate of Bacteroides fragilis in Scandinavia: genetic analysis of a unique insertion sequence. *Scand. J. Infect. Dis. 37*, 676-679.

Warren,R.M., Sampson,S.L., Richardson,M., van der Spuy,G.D., Lombard,C.J., Victor,T.C., and van Helden,P.D. (2000). Mapping of IS*6110* flanking regions in clinical isolates of *M. tuberculosis* demonstrates genome plasticity. *Mol. Microbiol. 37*, 1405-1416.

Webber,M.A. and Piddock,L.J. (2001). Absence of mutations in marRAB or soxRS in acrB-overexpressing fluoroquinolone-resistant clinical and veterinary isolates of Escherichia coli. *Antimicrob. Agents Chemother. 45*, 1550-1552.

Willems,R.J., Top,J., van den,B.N., van,B.A., Mevius,D.J., Hendriks,G., van Santen-Verheuvel,M., and van Embden,J.D. (1999). Molecular diversity and evolutionary relationships of Tn1546-like elements in enterococci from humans and animals. Antimicrob. *Agents Chemother. 43*, 483-491.

Zafarullah,M., Charlier,D., and Glansdorff,N. (1981). Insertion of IS3 can "turn-on" a silent gene in Escherichia coli. *J. Bacteriol. 146*, 415-417.

Zhou,H., Pi,B.R., Yang,Q., Yu,Y.S., Chen,Y.G., Li,L.J., and Zheng,S.S. (2007). Dissemination of imipenem-resistant Acinetobacter baumannii strains carrying the ISAba1 blaOXA-23 genes in a Chinese hospital. *J. Med. Microbiol. 56*, 1076-1080.

Zuerner,R.L. and Trueba,G.A. (2005). Characterization of IS1501 mutants of Leptospira interrogans serovar pomona. *FEMS Microbiol. Lett. 248*, 199-205.

In: DNA Transposable Elements Research
Editors: K. Yoshida, M. Aoki, pp. 157-165
ISBN: 978-1-60456-531-7
© 2008 Nova Science Publishers, Inc.

Chapter VI

Transposon-Based Technologies in the Ascidian *Ciona Intestinalis*

Yasunori Sasakura
Shimoda Marine Research Center, University of Tsukuba, 5-10-1, Shimoda, Shizuoka, 415-0025, Japan

Abstract

Ascidians, which are commonly known as "sea squirts", belong to phylum Chordata with cephalochordates and vertebrates. As this phylogenetic position suggests, ascidians share characteristics with vertebrates, such as dorsally located neural tubes and notochords. A cosmopolitan species of ascidian, *Ciona intestinalis*, is commonly used as a material for developmental and genome biology. The draft genome sequences of *Ciona* have revealed that this ascidian possesses a compact genome with little redundancy of genes. The *Ciona* life cycle is approximately two to three months, and inland culture systems have been developed. Thus, *Ciona* provides an excellent material on which to perform genetic analyses. Recent studies have shown that a Tc1/*mariner* superfamily transposon *Minos* causes both excision and transposition in *Ciona*. Activity of *Minos* is sufficient to cause germline transformation. So far, techniques such as creation of marker transgenic lines, enhancer detection and insertional mutagenesis have been introduced in *Ciona* with the *Minos* system. *Minos* inserted in the *Ciona* genome can be remobilized by providing transposase in the transgenic lines and, utilizing the remobilization technique, efficient enhancer detection screening has been established. In this article, these recent technological achievements of *Ciona* genetics based on the *Minos* transposon are reviewed.

Introduction

Transposon-based technologies are now widely available for studying biological processes. In contrast to their usefulness, however, transposon technologies have been introduced in only a few organisms. One major obstacle to a wider introduction is the difficulty in finding a transposon that is active in an organism, mainly because the mechanisms that determine transposon activity are not well understood. For example, transposons are harmful to genomes because transposon insertions disrupt gene functions, and their activity is usually repressed to protect an organism, but knowledge about such repressive mechanisms is fragmental. For instance, researchers have not yet found a hyperactive transposon in all organisms. Accurate prediction of whether a well-studied transposon is active in an organism is also impossible. Researchers must test transposon activity one by one by laborious experimenting. In such circumstances, the identification of a transposon active in various organisms is quite important.

Recently it was discovered that a Tc1/*mariner* superfamily transposon, *Minos*, is active in a marine invertebrate chordate, *Ciona intestinalis*, enough to cause germline transformation [1-5]. This discovery is the first report on transposon activity in marine invertebrates. With *Minos*, transposon techniques such as the creation of stable transgenic lines, enhancer detection and insertional mutagenesis have been introduced in *Ciona intestinalis* [6, 7]. Based on this achievement, *Minos* has been used in various organisms and its consistent activity has been shown. In this chapter, *Ciona intestinalis* as a new material for genetics study and transposon technologies in *Ciona* and the *Minos* transposon as a powerful transformation tool are reviewed.

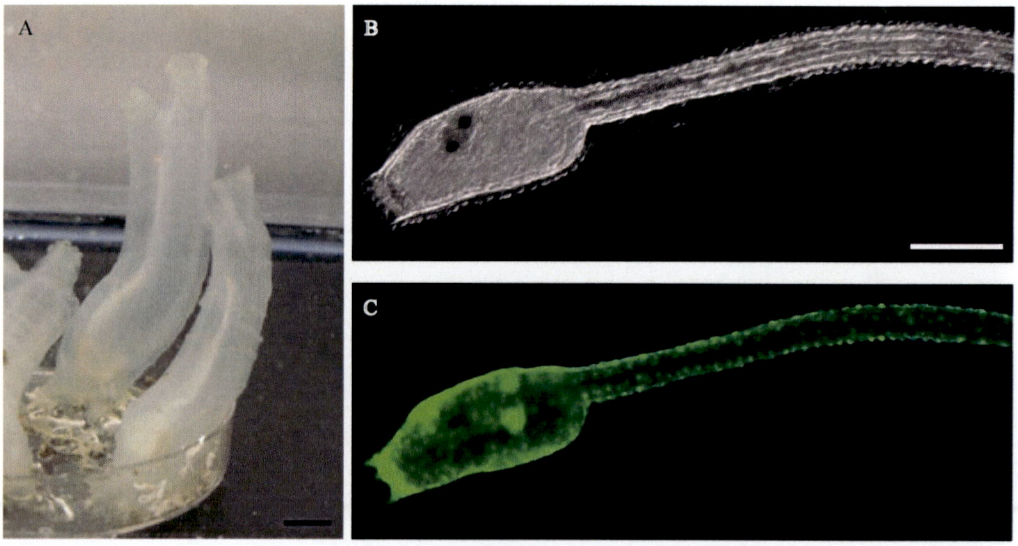

Figure 1 (A) Adults of *Ciona intestinalis*. *Ciona* adults are sessile animals; they adhere to substrate at the posterior side of the body. Bar, 10 mm. (B) A swimming larva of *Ciona intestinalis*. Ascidian larvae are typical tadpole larvae, and they swim vigorously with tail. Bar, 100 μm. (C) GFP expression in a larva of a stable transgenic line that was created with *Minos* transposon. GFP was expressed in epidermis.

Ciona Intestinalis as a Material for Genetics Study

The ascidian *Ciona intestinalis* (hereafter referred to as "*Ciona*") provides a splendid experimental system for genetics study [5, 8]. Its life cycle is approximately 2-3 months, which is short enough to enable us to culture *Ciona* in laboratories. There are several established systems of *Ciona* inland culture [9]. These closed culture systems make it possible to maintain *Ciona* transgenic lines without spreading them in the ocean. Several laboratories are preparing to become stock centers for *Ciona* transgenic lines and mutants [9].

A major difference in *Ciona* genetics compared with other animals is this species' sex differentiation. Ascidians including *Ciona* are hermaphroditic, and *Ciona* gametes are self-fertile when they are obtained by spawning [4, 10]. The ability of self-fertilization is advantageous for mutant screening. Transposon-mediated mutagenesis requires characterization of animals having the same insertions by molecular techniques such as Southern blotting to create homozygous animals with respect to the insertion sites [11], and this step is time-consuming. If progeny can be generated with self-fertilization, the progeny contain homozygous animals with respect to each insertion site, irrespective of the number of insertions carried by their single parent. Therefore, we can reduce the time and cost associated with the identification of insertion sites. Surgically isolated gametes do not have a self-fertilization ability. Therefore, gametes with or without self-fertilization can be obtained.

The outstanding advantage of *Ciona* over other ascidians is the availability of the draft genome sequences [12]. The genome project revealed that *Ciona* has a compact genome. The genome size is approximately 160 Mbp per haploid, with 15,852 protein-coding genes. The genome size and gene number is similar to that of *Drosophila melanogaster* and much smaller than that of most vertebrates. In addition, the *Ciona* gene set is basically non-redundant compared with that of vertebrates [13-16]. In vertebrate genomes, there are several closely related genes whose sequences are very similar. Such paralogous genes often compensate for their functions, and a knockdown of one of them does not show severe phenotypes. In many cases, the *Ciona* genome contains a single orthologous gene, and disruption of the gene may be sufficient to analyze its functions. The compact and non-redundant genome assures frequent mutagenesis. Thus, *Ciona* is a splendid material for genetic experiments.

Transposon Technologies in *Ciona Intestinalis*

In *Ciona intestinalis,* a transposon *Minos* has activity sufficient to perform germline transformation [1, 2]. *Minos* is a member of the Tc1/*mariner* transposon superfamily [17]. Tc1/*mariner* transposons comprise a famous DNA transposon family and are utilized as tools for germline transformation and insertional mutagenesis in various animals [18, 19]. Tc1/*mariner* transposons are small and simple. This transposon contains an open reading frame (ORF) that encodes a transposase, and the ORF is flanked by two inverted repeats [17, 20-24]. The ORF can be replaced with artificial DNA fragments without loss of transposon activity. Tc1/*mariner* transposons are inserted in DNA by targeting TA dinucleotides. The simple target sequences indicate that the transposons can be inserted into genomes without strong hot spots. In the case of Tc3 and *Himar*1, their recombinant transposases purified from *E. coli* can cause *in vitro* transposition [25, 26]. Because the jumping mechanisms of Tc1/*mariner* transposons are similar, the reports suggest that this transposon family does not require a co-factor for transposition. This explains why Tc1/*mariner* transposons show transposition activities in various non-host organisms. Several studies have shown that another Tc1/*mariner* transposon, *Sleeping Beauty* (*SB*), requires co-factors [27, 28]. The contradiction might suggest that transposition mechanisms of Tc1/*mariner* transposons are similar but have different points among transposons. In spite of this contradiction, Tc1/*mariner* transposons are excellent transposons as tools for germline transformation because of their loose host specificity.

Minos was originally isolated from *Drosophila hydei* [17], and it shows transposition activity in various animals, including the fruit fly *Drosophila melanogaster*, the olive fly *Bactrocera oleae*, the mosquitos *Aedes aegypti* and *Anopheles stephensi*, the silk worm *Bombyx mori*, the beetle *Tribolium castaneum*, the crustacean *Parhyale hawaiensis*, the basal chordate *Ciona intestinalis* and its related species *C. savignyi*, the vertebrates mouse and human HeLa cells [1, 2, 29-46]. In *Ciona intestinalis*, founder animals were created by microinjection or electroporation of a plasmid DNA containing *Minos* and *in vitro* synthesized transposase mRNA [2, 3]. About 30% of transposon- and transposase-introduced *Ciona* individuals become founders and transmit insertions to the subsequent generations (Sasakura et al., 2003; Matsuoka et al., 2005). The insertion sites can be determined by thermal asymmetric interlaced (TAIL)-PCR [47]. With *Minos*, several transgenic lines that express a reporter gene in a tissue-specific manner have been established [9]. In addition, analyses of a *Minos* insertional mutant, *swimming juvenile*, have shown that ascidian cellulose synthase has critical functions in cellulose biosynthesis and is necessary for normal metamorphosis [7].

So far, the transformation of *Ciona* with other transposons has not been reported. The excision activities of *SB* and *Frog Prince* (*FP*) have been described in [5]. *SB* and *FP* show excision in *Ciona*. However, their activities are inefficient, and germline transformation with these transposons has failed. This is in contrast to the case with vertebrates, in which *SB* and *FP* are reliable transposons to cause efficient germline transformation [23, 24, 48-50]. Investigation of the mechanisms that restrict the activities of *SB* and *FP* in *Ciona* will help us understand the transposition mechanisms of Tc1/*mariner* transposons in non-host organisms.

Enhancer Detection in *Ciona Intestinalis*

Enhancer detection is one of the most famous applications of transposons [51, 52]. Generally, a transposon construct that contains a reporter gene with a weak promoter is used for this technology. In many cases, such transposon constructs show weak or no reporter gene expression. When the transposons are inserted into a genomic region under the effect of an enhancer, the promoter in the transposon is controlled by the enhancer. The efficiency of reporter gene expression is enhanced, and the pattern is altered. Usually a reporter gene shows tissue-specific expressions according to the enhancer activities and thus "entraps" the enhancers.

Ciona is a suitable material for enhancer detection. Its compact genome and high density of genes in the genome suggest the presence of many enhancers in the genome. In *Ciona*, enhancer detection with a promoter of *Ci-TPO*, which encodes thyroid peroxidase orthologue, and with green fluorescent protein (GFP) has been reported [6]. It was estimated that about 5.4% of *Minos*-introduced animals inherit enhancer detection insertions [4, 5]. Many enhancer detection lines have been created that show tissue-specific GFP expression in such tissues as epidermis, muscle, endostyle, pharyngeal gill and digestive organs. These transgenic lines are useful marker lines to analyze tissue formation as well as identify tissue-specific enhancers.

The advantage of *Ciona*'s compact genome and high frequency enhancer detection are that genome-wide enhancer detection is probable in *Ciona*. For this purpose, a way to perform highly efficient enhancer detection is necessary, and such a method has been achieved in *Ciona*, which utilizes re-mobilization of *Minos* in the *Ciona* genome [53]. In this method, transposase is expressed in a transgenic line that has insertions of *Minos* for enhancer detection. The expressed transposases cut and paste *Minos* in the *Ciona* genome to create new insertions. When transposase mRNA was provided by microinjection, as many as 80% of transposase-injected animals inherit enhancer detection insertions to pass on to subsequent generations (Awazu et al., 2007).

Transposase-expressing lines, so-called "jump-starter" lines, have been created in the *Ciona Minos* system [54]. These lines express *Minos* transposase in sperm. New transposon insertions can be created from double transgenic animals between transposase-expressing lines and transposon insertion lines. These transposase-expressing lines were applied to enhancer detection. All double-transgenic animals between a transposase-expressing line and a transposon insertion line that had transposon vectors for enhancer detection passed several enhancer detection events to their progeny. High-frequency enhancer detection helps to isolate various enhancer detection lines on a restricted time schedule, thus facilitating genome-wide enhancer detection.

Conclusion

The *Minos* transposon provides us with an efficient transformation vector in various organisms. With *Minos*, techniques of developmental genetics have been introduced in an ascidian, *Ciona intestinalis*. By this achievement, *Ciona* is regarded as being on the frontier of developmental genetics in marine invertebrates; the success in *Ciona* suggests that genetic approaches can be introduced in other marine invertebrates with *Minos*.

Acknowledgments

I thank all members of the Shimoda Marine Research Center, University of Tsukuba, as well as Prof. Charalambos Savakis (Crete University, Greece), Dr. Koichi Kawakami

(National Institute of Genetics, Japan) and Prof. Nori Satoh (Kyoto University, Japan) for their kind help with my research. This study was supported by Grants-in-Aid for Scientific Research from JSPS and MEXT, NIG Cooperative Research Program (2007-B01).

References

[1] Sasakura, Y., Awazu, S., Chiba, S., Kano, S. and Satoh, N. (2003). Application of *Minos*, one of the Tc1/*mariner* superfamily transposable elements, to ascidian embryos as a tool for insertional mutagenesis. *Gene* 308, 11-20.

[2] Sasakura, Y., Awazu, S., Chiba, S. and Satoh, N. (2003). Germ-line transgenesis of the Tc1/*mariner* superfamily transposon *Minos* in *Ciona intestinalis*. *Proc. Natl. Acad. Sci. USA* 100, 7726-7730.

[3] Matsuoka, T., Awazu, S., Shoguchi, E., Satoh, N. and Sasakura, Y. (2005). Germline transgenesis of the ascidian *Ciona intestinalis* by electroporation. *Genesis* 41, 61-72.

[4] Sasakura, Y. (2007). Germline transgenesis and insertional mutagenesis in the ascidian *Ciona intestinalis*. *Dev. Dyn.* 236, 1758-1767.

[5] Sasakura, Y., Oogai, Y., Matsuoka, T., Satoh, N. and Awazu, S. (2007). Transposon-mediated transgenesis in a marine invertebrate chordate, *Ciona intestinalis*. *Genome Biol.* 8, S3.

[6] Awazu, S., Sasaki, A., Matsuoka, T., Satoh, N. and Sasakura, Y. (2004). An enhancer trap in the ascidian *Ciona intestinalis* identifies enhancers of its *Musashi* orthologous gene. *Dev. Biol.* 275, 459-472.

[7] Sasakura, Y., Nakashima, K., Awazu, S., Matsuoka, T., Nakayama, A., Azuma, J. and Satoh, N. (2005). Transposon-mediated insertional mutagenesis revealed the functions of animal cellulose synthase in the ascidian *Ciona intestinalis*. *Proc. Natl. Acad. Sci. USA* 102, 15134-15139.

[8] Sordino, P., Belluzzi, L., De Santis, R. and Smith, W. C. (2001). Developmental genetics in primitive chordates. Philos. Trans. R. Soc. Lond. *B. Biol. Sci.* 356, 1573-1582.

[9] Joly, J. S., Kano, S., Matsuoka, T., Auger, H., Hirayama, K., Satoh, N., Awazu, S., Legendre, L. and Sasakura, Y. (2007). Culture of *Ciona intestinalis* in closed systems. *Dev. Dyn.* 236, 1832-1840.

[10] Jiang, D. and Smith, W. C. (2005). Self- and cross-fertilization in the solitary ascidian *Ciona savignyi*. *Biol. Bull.* 209, 107-112.

[11] Amsterdam, A. and Hopkins, N. (2004). Retrovirus-mediated insertional mutagenesis in zebrafish. *Methods Cell Biol.* 77, 3-20.

[12] Dehal, P., Satou, Y., Campbell, R.K., Chapman, J., Degnan, B., De Tomaso, A., Davidson, B., Di Gregorio, A., Gelpke, M., Goodstein, D.M., Harafuji, N., Hastings, K.E., Ho, I., Hotta, K., Huang, W., Kawashima, T., Lemaire, P., Martinez, D., Meinertzhagen, I.A., Necula, S., Nonaka, M., Putnam, N., Rash, S., Saiga, H., Satake, M., Terry, A., Yamada, L., Wang, H.G., Awazu, S., Azumi, K., Boore, J., Branno, M., Chin-Bow, S., DeSantis, R., Doyle, S., Francino, P., Keys, D.N., Haga, S., Hayashi, H., Hino, K., Imai, K.S., Inaba, K., Kano, S., Kobayashi, K., Kobayashi, M., Lee, B.I.,

Makabe, K.W., Manohar, C., Matassi, G., Medina, M., Mochizuki, Y., Mount, S., Morishita, T., Miura, S., Nakayama, A., Nishizaka, S., Nomoto, H., Ohta, F., Oishi, K., Rigoutsos, I., Sano, M., Sasaki, A., Sasakura, Y., Shoguchi, E., Shin-I, T., Spagnuolo, A., Stainier, D., Suzuki, M.M., Tassy, O., Takatori, N., Tokuoka, M., Yagi, K., Yoshizaki, F., Wada, S., Zhang, C., Hyatt, P.D., Larimer, F., Detter, C., Doggett, N., Glavina, T., Hawkins, T., Richardson, P., Lucas, S., Kohara, Y., Levine, M., Satoh, N. and Rokhsar, D.S. (2002). The draft genome of *Ciona intestinalis*: Insight into chordate and vertebrate origins. *Science* 298, 2157-2167.

[13] Sasakura, Y., Shoguchi, E., Takatori, N., Wada, S., Meinertzhagen, I. A., Satou, Y. and Satoh, N. (2003). A genomewide survey of developmentally relevant genes in *Ciona intestinalis*. X. Genes for cell junctions and extracellular matrix. *Dev. Genes Evol.* 213, 303-313.

[14] Sasakura, Y., Yamada, L., Takatori, N., Satou, Y. and Satoh, N. (2003). A genomewide survey of developmentally relevant genes in *Ciona intestinalis*. VII. Molecules involved in the regulation of cell polarity and actin dynamics. *Dev. Genes Evol.* 213, 273-283.

[15] Satou, Y., Sasakura, Y., Yamada, L., Imai, K. S., Satoh, N. and Degnan, B. (2003). A genomewide survey of developmentally relevant genes in *Ciona intestinalis* V. Genes for receptor tyrosine kinase pathway and Notch signaling pathway. *Dev. Genes Evol.* 213, 254-263.

[16] Satou, Y. and Satoh, N. (2003). Genomewide surveys of developmentally relevant genes in *Ciona intestinalis*. *Dev. Genes Evol.* 213, 211-213.

[17] Franz, G. and Savakis, C. C. (1991). *Minos*, a new transposable element from *Drosophila hydei*, is a member of the *Tc1*-like family of transposons. . *Nucleic Acid Res.* 19, 6646.

[18] Plasterk, R., Izsvak, Z. and Ivics, Z. (1999). Resident aliens: The Tc1/*mariner* superfamily of transposable elements. *Trends Genet* 15, 326-332.

[19] Miskey, C., Izsvak, Z., Kawakami, K. and Ivics, Z. (2005). DNA transposons in vertebrate functional genomics. *Cell Mol. Life Sci.* 62, 629-641.

[20] Rosenzweig, B., Liao, L. W. and Hirsh, D. (1983). Sequence of the *C. elegans* transposable element Tc1. *Nucleic Acids Res.* 11, 4201-4210.

[21] Jacobson, J. W., Medhora, M. M. and Hartl, D. L. (1986). Molecular structure of a somatically unstable transposable element in Drosophila. *Proc. Natl. Acad. Sci. USA* 83, 8684-8688.

[22] Plasterk, R. H. A. and van Luenen, H. G. A. M. In *C. elegans* II Riddle, D. L., Blumenthal, T., Meyer, B. J. and Priess, J. R. Ed. Cold Spring Harbor Laboratory Press, New York, NY, 1996, pp 97-116.

[23] Ivics, Z., Plasterk, R. H., Hackett, P. B. and Izsvak, Z. (1997). Molecular reconstruction of *Sleeping Beauty*, a *Tc1*-like transposon from fish, and its transposition in human cells. *Cell* 91, 501-510.

[24] Miskey, C., Izsvak, Z., Plasterk, R. H. and Ivics, Z. (2003). The *Frog Prince*: a reconstructed transposon from *Rana pipiens* with high transpositional activity in vertebrate cells. *Nucleic Acid Res.* 31, 6873-6881.

[25] Lampe, D., Churchill, M. E. A. and Robertson, H. M. (1996). A purified mariner transposase is sufficient to mediate transposition *in vitro*. *EMBO J.* 15, 5470-5479.

[26] Vos, J. C., De Baere, I. and Plasterk, R. H. (1996). Tc1 transposase is the only nematode protein required for *in vitro* transposition of Tc1. *Genes Dev.* 10, 755-761.

[27] Zayed, H., Izsvak, Z., Khare, D., Heinemann, U. and Ivics, Z. (2003). The DNA-bending protein HMGB1 is a cellular cofactor of Sleeping Beauty transposition. *Nucleic Acids Res.* 31, 2313-2322.

[28] Walisko, O., Izsvák, Z., Szabo, K., Kaufman, C. D., Herold, S. and Ivics, Z. (2006). *Sleeping Beauty* transposase modulates cell-cycle progression through interaction with Miz-1. *Proc. Natl. Acad. Sci. USA* 103, 4062-4067.

[29] Loukeris, T. G., Arca, B., Livadaras, I., Dialektaki, G. and Savakis, C. (1995). Introduction of the transposable element *Minos* into the germ line of *Drosophila melanogaster*. *Proc. Natl. Acad. Sci. USA* 92, 9485-9489.

[30] Loukeris, T. G., Livadaras, I., Arca, B., Zabulou, S. and Savakis, C. (1995). Gene transfer into the Medfly, *Ceratitis capitata*, with a *Drosophila hydei* transposable element. *Science* 270, 2002-2005.

[31] Catteruccia, F., Nolan, T., Loukeris, T. G., Blass, C., Savakis, C., Kafatos, F. C. and Crisanti, A. (2000). Stable germline transformation of the malaria mosquito *Anopheles stephensi*. *Nature* 405, 959-962.

[32] Catteruccia, F., Nolan, T., Blass, C., Muller, H. M., Crisanti, A., Kafatos, F. C. and Loukeris, T. G. (2000). Toward Anopheles transformation: Minos element activity in anopheline cells and embryos. *Proc. Natl. Acad. Sci. USA* 97, 2157-2162.

[33] Klinakis, A. G., Loukeris, T. G., Pavlopoulos, A. and Savakis, C. (2000). Mobility assays confirm the broad host-range activity of the *Minos* transposable element and validate new transformation tools. *Insect Mol. Biol.* 9, 269-275.

[34] Klinakis, A. G., Zagoraiou, L., Vassilatis, D. K. and Savakis, C. (2000). Genome-wide insertional mutagenesis in human cells by the *Drosophila* mobile element *Minos*. *EMBO reports* 1, 416-421.

[35] Shimizu, K., Kamba, M., Sonobe, H., Kanda, T., Klinakis, A. G., Savakis, C. and Tamura, T. (2000). Extrachromosomal transposition of the transposable element *Minos* in embryos of the silkworm *Bombyx mori*. *Insect Mol. Biol.* 9, 277-281.

[36] Kapetanaki, M. G., Loukeris, T. G., Livadaras, I. and Savakis, C. (2002). High frequencies of *Minos* transposon mobilization are obtained in insects by using *in vitro* synthesized mRNA as a source of transposase. *Nucleic Acid Res.* 30, 3333-3340.

[37] Zhang, H., Shinmyo, Y., Hirose, A., Inoue, Y., Ohuchi, H., Loukeris, T., Eggleston, P. and Noji, S. (2002). Extrachromosomal transposition of the transposable element *Minos* in embryos of the cricket *Gryllus bimaculatus*. Develop. *Growth Differ.* 44, 409-417.

[38] Drabek, D., Zagoraiou, L., deWit, T., Langeveld, A., Roumpaki, C., Mamalaki, C., Savakis, C. and Grosveld, F. (2003). Transposition of the *Drosophila hydei Minos* transposon in the mouse germ line. *Genomics* 81, 108-111.

[39] Matsuoka, T., Awazu, S., Satoh, N. and Sasakura, Y. (2004). *Minos* transposon causes germline trasngenesis of the ascidian *Ciona savignyi*. Dev. Growth Differ. 46, 249-255.

[40] Pavlopoulos, A., Berghammer, A. J., Averof, M. and Klingler, M. (2004). Efficient transformation of the beetle *Tribolium castaneum* using *Minos* transposable element:

quantitative and qualitative analysis of genomic integration events. *Genetics* 167, 737-746.

[41] Metaxakis, A., Oehler, S., Klinakis, A. and Savakis, C. (2005). *Minos* as a genetic and genomic tool in *Drosophila melanogaster. Genetics* 171, 571-581.

[42] Pavlopoulos, A. and Averof, M. (2005). Establishing genetic transformation for comparative developmental studies in the crustacean *Parhyale hawaiensis. Proc. Natl. Acad. Sci. USA* 102, 7888-7893.

[43] Koukidou, M., Klinakis, A., Reboulakis, C., Zagoraiou, L., Tavernarakis, N., Livadaras, I., Economopoulos, A. and Savakis, C. (2006). Germ line transformation of the olive fly *Bactrocera oleae* using a versatile transgenesis marker. *Insect Mol. Biol.* 15, 95-103.

[44] Pavlopoulos, A., Oehler, S., Kapetanaki, M. G. and Savakis, C. (2007). The DNA transposon *Minos* as a tool for transgenesis and functional genomic analysis in vertebrates and invertebrates. *Genome Biol.* 8, S2.

[45] Scali, C., Nolan, T., Sharakhov, I., Sharakhova, M., Crisanti, A. and Catteruccia, F. (2007). Post-integration behavior of a *Minos* transposon in the malaria mosquito *Anopheles stephensi.* Mol. Genet. *Genomics* 278, 575-584.

[46] Uchino, K., Imamura, M., Shimizu, K., Kanda, T. and Tamura, T. (2007). Germ line transformation of the silkworm, Bombyx mori, using the transposable element *Minos.* Mol. *Genet. Genomics* 277, 213-220.

[47] Liu, Y. G., Mitsukawa, N., Oosumi, T. and Whittier, R. F. (1995). Efficient isolation and mapping of *Arabidopsis thaliana* T-DNA insert junctions by thermal asymmetric interlaced PCR. *Plant J.* 8, 457-463.

[48] Fischer, S. E., Wienholds, E. and Plasterk, R. H. A. (2001). Regulated transposition of a fish transposon in the mouse germ line. *Proc Natl Acad Sci USA* 98, 6759-6764.

[49] Davidson, A. E., Balciunas, D., Mohn, D., Shaffer, J., Hermanson, S., Sivasubbu, S., Pat Cliff, M., Hackett, P. B. and Ekker, S. C. (2003). *Efficient gene delivery and* gene expression in zebrafish using the *Sleeping Beauty* transposon. Dev. Biol. 263, 191-202.

[50] Sinzelle, L., Vallin, J., Coen, L., Chesneau, A., Du Pasquier, D., Pollet, N., Demeneix, B. and Mazabraud, A. (2006). Generation of transgenic *Xenopus laevis* using the *Sleeping Beauty* transposon system. *Transgenic Res.* 15, 751-760.

[51] O'Kane, C. and Gehring, W. J. (1987). Detection in situ of genomic regulatory elements in *Drosophila. Proc. Natl. Acad. Sci. USA* 84, 9123-9127.

[52] Bellen, H. J. (1999). Ten years of enhancer detection: lessons from the fly. *The Plant Cell* 11, 2271-2281.

[53] Awazu, S., Matsuoka, T., Inaba, K., Satoh, N. and Sasakura, Y. (2007). High-throughput enhancer trap by remobilization of transposon *Minos* in *Ciona intestinalis. Genesis* 45, 307-317.

[54] Sasakura, Y., Konno, A., Mizuno, K., Satoh, N. and Inaba, K. (2008). Enhancer detection in the ascidian *Ciona intestinalis* with transposase-expressing lines of *Minos. Dev. Dyn.* 237, 39-50.

In: DNA Transposable Elements Research
Editors: K. Yoshida, M. Aoki, pp. 167-179
ISBN: 978-1-60456-531-7
© 2008 Nova Science Publishers, Inc.

Chapter VII

CRISPR Loci in *Corynebacterium Diphtheriae*: *In Silico* Analysis and Insights into Their Origin, Evolution and Function

*Igor Mokrousov**
Laboratory of Molecular Microbiology, St. Petersburg Pasteur Institute, St. Petersburg, Russia

Abstract

CRISPRs (clustered, regularly interspaced short palindromic repeats) are loci present in many bacterial genomes and characterized by alternating minisatellite direct repeats (DR) and non-repeats (spacers). In the complete genome sequence of *Corynebacterium diphtheriae* strain NCTC13129, we previously identified two CRISPR loci 39 kb downstream and 180 kb upstream of the *oriC* region, that we named DRA and DRB, respectively. Here, we analysed these loci in silico in more detail. A search against complete GenBank revealed a homology of seven spacers in both loci (one DRA spacer and six DRB spacers). Noteworthy, four of them showed such homology with either phage hypothetical proteins, or plasmid transposon encoded genes related to integration or recombination. This result corroborates with recent observations made on other bacterial genomes, suggesting that CRISPRs are structures able to take up pieces of foreign DNA as part of a defence mechanism against its invasion. In *C. diphtheriae*, five of the seven found homologies are on minus-strand sequences in GenBank hinting at the RNA interference as a mode of action in CRISPR loci. CRISPR analysis also revealed striking points of similarity between *C. diphtheriae* and other *Corynebacterium* (*C. jeikeium*) partly owing to the horizontal gene transfer. In particular, the last spacer in the *C. diphtheriae* DRA locus showed homology to the direct repeat motif in the large 60-unit CRISPR locus in *C. jeikeium*. *C. jeikeium* is a skin pathogen and thus likely a more ancient species than human respiratory pathogen *C. diphtheriae*. Accordingly, it seems more plausible that this spacer was acquired from *C. jeikeium* CRISPR locus and not *vice versa*. Analysis of the regions flanking both CRISPR loci in *C. diphtheriae* suggests that at least one them (DRA), functionally or structurally, may be prone to acquire DNA fragments, acting as "insertion preferential locus" (*ipl*).

Introduction

Repetitive genome sequences present important source of intraspecies variation. A novel family of such loci (clustered, regularly interspaced short palindromic repeats [CRISPR]) has recently been identified by in silico analysis in many bacterial species (Jansen et al. 2002). This family is characterized by direct repeats varying in size from 21 to 37 bp, interspaced by similarly sized non-repetitive sequences (variable spacers). CRISPRs are intriguing regions in bacterial genomes. Their origin and biological function long remained unclear. Recent studies based on use of experimental models and in silico analysis gave new insights into origin and evolution of CRISPRs. It has been demonstrated that their spacers are not unique, as previously considered (Jansen et al. 2002; Mojica et al., 2000), but derive from preexisting sequences from closely related but not the same species, mainly though not exclusively from extrachromosomal elements (Bolotin et al. 2005; Mojica et al. 2005; Pourcel et al. 2005; Saunders et al. 2005). It has been hypothesized that spacer elements are the traces of past invasions, providing the cell immunity against phage infection, and more generally foreign DNA expression, by coding an anti-sense RNA. The following rules may define CRISPR evolution: (i) random deletions of single spacers or contiguous blocks may occur; (ii) in contrast, the addition of new motifs is polarized and requires that the last constant region near the leader sequence is duplicated and that a piece of DNA about 30 bp in length (the spacer) is simultaneously copied and added; (iii) the presence of identical spacers in a CRISPR allele reflects shared ancestry and does not result from independent events (van Embden et al. 2000; Pourcel et al. 2005). In *Yersinia pestis*, addition of spacers was shown to be an ongoing process (Pourcel et al. 2005). Contrarily, the addition of new spacers has not been observed in *Mycobacterium tuberculosis* (van Embden et al. 2000) and the presence of insertion elements may indicate that this locus in *M. tuberculosis* is no more active.

In the recently published complete genome sequence of *Corynebacterium diphtheriae* strain NCTC13129 (Cerdeño-Tarraga et al. 2003), we previously identified *in silico* two CRISPR loci 39 kb downstream and 180 kb upstream of the *oriC* region, consisting of minisatellite (27-36 bp) alternating direct repeats and variable spacers (Mokrousov et al. 2005). We designated these loci as DRA and DRB, respectively. In the present study, we analysed in more detail both CRISPR loci in *C. diphtheriae*, attempting to gain new insights into their evolution and function.

Methods

Genome search for repeated sequences in the complete genome sequence of *C. diphtheriae* strain NCTC13129 (Genbank accession number NC_002935) was done using Tandem Repeat Finder software (2). Settings used were: alignment parameters (match, mismatch, indel) +2, -3, -5; maximum period size 100, minimum alignment score 30. The obtained hits were manually searched for presence of multiple, short (50-80 bp unit size), non-exact (homology 50-80%) repeats. This analysis identified location and structure of the two regions that corresponded to the definition of the DR/CRISPR loci (Jansen et al. 2002). They are situated downstream and upstream of the origin of replication (*oriC*), and we designated them as DRA and DRB loci, respectively (Figure 1). The sequences of these two loci are shown in Figures 2 and 3. For convenience, spacers of the DRA and DRB loci were designated by adding 'a' and 'b', respectively, to the spacer number.

BLAST nucleotide search engine (www.ncbi.nlm.nih.gov/BLAST) was used for sequence search in the GenBank. Phylogenetic analysis of the aligned nucleotide sequences was done with MEGA 2.0 software (Kumar et al. 2001).

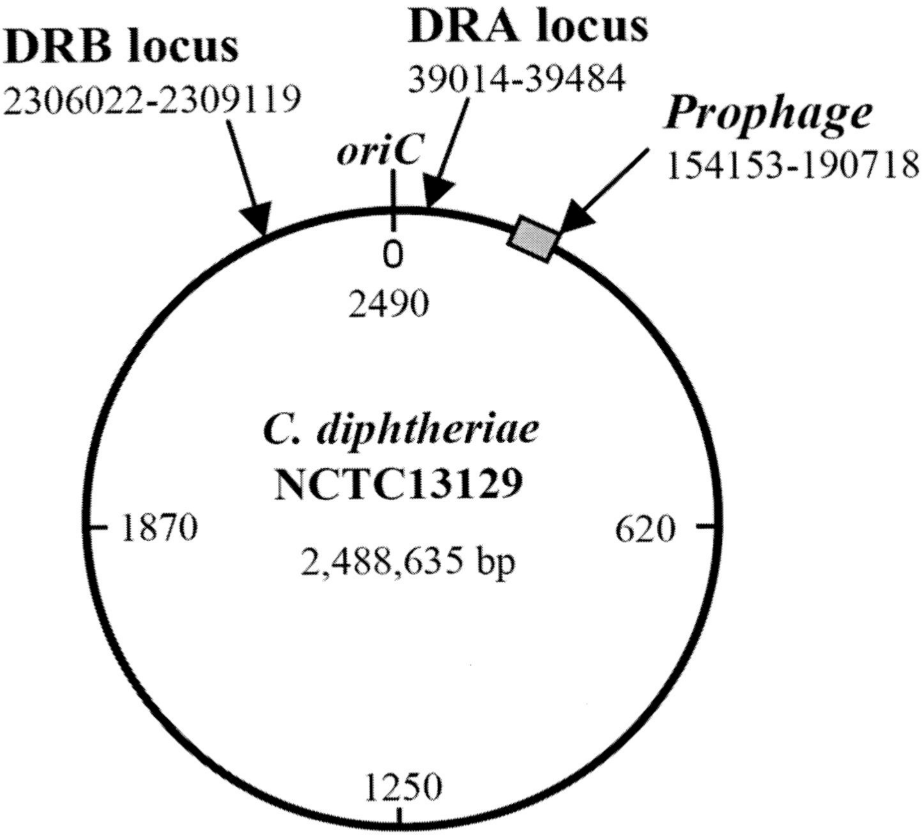

Figure 1. Location of the CRISPR loci in the *C. diphtheriae* genome.

```
DRA locus (39014-)                                              0a
gaagtctatcagggttttgagaactgaaccccagtgcggttcaacactcaaaaacaattccatgaagtctatcagggtt

                                        1a
tttgagaactgaaccccagtacgctgcctgagccttgagcgagtgctggaagtctatcagggttttgagaactgaaccc

                 2a                                                             3a
cagttgcgggctttggtggtgtcggcgtccacgaagtctatcagggttttgagaactgaaccccagtgcgcgagcggct

                                                                      4a
tccacgtaccatagaagaagtctatcagggttttgagaactgaaccccagtatcgtctaacagttgctaaatttttgc

                                                      5a
gaagtctatcagggttttgagaactgaaccccagttcggcgcgagcctcgctaggtgctgctggaagtctatcagggtt

                                        6a
tttgagaactgaaccccagcccgcgccagcggagatgagccgtaggagaagtttatcagggttttgagaa (39484)
```

Figure 2. DRA CRISPR locus in *Corynebacterium diphtheriae* NCTC13129, complete genome (Cerdeno-Tarraga et al., 2003) positions from 39014 to 39484 (in relation to *oriC*), accession number BX248354 BX248353. DRs are underlined (last repeat is truncated), spacer number is given above the spacer sequence.

Results and Discussion

Recent reports, including extensive *in silico* analysis (Mojica et al. 2005) did not involve *C. diphtheriae*, a human respiratory pathogen that dramatically and epidemically re-emerged in the 1990s in the countries of the former Soviet Union (140,000 cases, 4,000 deaths in 1991-1996 [Vitek and Wharton 1998]) while imported cases were also registered in Europe (Popovic et al. 2000). We searched spacer sequences from both *C. diphtheriae* CRISPR loci against complete genome sequence of *C. diphtheriae* strain NCTC13129. As a result, no significant homology was found. Interestingly, although there is a corynephage in *C. diphtheriae* genome in a relative proximity to the DRA locus (Figure 1), it does not have homology with spacers. However, a further search against complete GenBank revealed similarity of seven spacers in both loci (one from DRA and six from DRB loci) (Table 1). Noteworthy, four of them showed such homology with either phage hypothetical proteins (6b, 12b), or plasmid transposon encoded genes related to integration or recombination (13b, 20b). This result corroborates with recent observations made on other bacterial genomes, both eukaryotic and archaeal (Bolotin et al. 2005; Mojica et al. 2005; Pourcel et al. 2005). It has been suggested that CRISPRs are structures able to take up pieces of foreign DNA as part of a defence mechanism, a kind of cell immunity against invasion of the foreign DNA. The preferential occurrence of CRISPR spacers derived from genetic elements that fail to infect the corresponding spacer carrier strain, but not from those successfully propagated in the population, strongly supports a relationship between CRISPR and such immunity (Bolotin et al. 2005; Mojica et al. 2005; Pourcel et al. 2005). The transcription of the CRISPR loci (Tang et al. 2002) suggests that such activity could be executed by CRISPR-RNA molecules, acting as regulatory RNA that specifically recognizes the target through the homologous RNA spacer sequence, similarly to the eukaryotic interference RNA.

Table 1. Homology search of the *C. diphtheriae* CRISPR spacers (DRA and DRB loci) in GenBank

Spacer	Spacer sequence (upper line) aligned to the matching sequence*	Source of the matching sequence, positions in accession number in GenBank and possible functionality
6a	CCGCGCCAGCGG**A**GATGAGCCGTAGGA CCGCGCCAGCGG**G**GATGAGCCGTA	*Corynebacterium jeikeium* K411 direct repeat motive in the large 60-unit CRISPR locus (total locus - 773367…777054 in CR931997)
6b	ATCGAAGTACCGGTCTGCCATGAGGTCGTTATA GAAGTACCGGTCTGCCAT	Stx1 converting bacteriophage (25586-25569 in AP005153) hypothetical protein Stx2 converting bacteriophage I (25587-25570 in AP005154) hypothetical protein Stx2 converting bacteriophage II (25685-25668 in AP004402) hypothetical protein Bacteriophage 933W (55600-55583 in AF125520) hypothetical protein Bacteriophage VT2-Sa (54641-54624 in AP000363) hypothetical protein *Escherichia coli* O157:H7 genomic DNA, Sakai-VT2 prophage inserted region (56955-56938 in AP000422) hypothetical protein
12b	GATGACCGCGGCCTTGTGCAGGTCAAGATTAG CGCGGCCTTGTGCAGGTC	Mycobacteriophage Che9c (24204-24187 in AY129333) hypothetical protein
13b	ATGGAAAGAAATCGCTGAACGCCGACTCCGCC GAAAGAAATCGCTGAAC	*Klebsiella pneumoniae* plasmid pRMH760 (4592-4576 in AY123253) Tn1696 TnpR resolvase *Klebsiella pneumoniae* NTUH-K2044 plasmid pK2044 (81422-81406 in AP006726) *Klebsiella pneumoniae* plasmid pLVPK (151749-151733 in AY378100) TnpR recombinase *Pseudomonas aeruginosa* plasmid R1033 transposon Tn1696 (3142-3126 in PAU12338) *Salmonella enterica* subsp. *enterica* serovar Choleraesuis plasmid pSC138 (8922-8906 in AY509004)Tn3926 TnpR resolvase family recombinase *Salmonella enterica* subsp. *enterica* serovar Typhi str. CT18 plasmid pHCM1 (122789-122805 in AL513383) putative membrane protein, resolvase *Salmonella typhimurium* virulence plasmid t-ST4, In-t7 and In-t8 integrons (11250-11234 n AJ746361) Tn1696 TnpR resolvase *Escherichia coli* partial *merA* gene, *merD* gene (2549-2565 in Y10102) *Enterobacter cloacae* DNA, mosaic mercury resistance transposable element (*mer*-operon) (4871-4887 in Y09025) TnpR resolvase *Alcaligenes sp.* pMER610 incomplete unit of mosaic mercury resistance transposon (5316-5332 in Y08993) *Yersinia enterocolitica* transposon TN3926 (1148-1164 in X78059) resolvase gene

Table 1 (Continued)

Spacer	Spacer sequence (upper line) aligned to the matching sequence*	Source of the matching sequence, positions in accession number in GenBank and possible functionality
15b	CACGTA**G**GCTTGTCGGGCCGCATATGAATGGT	
	CGTA**A**GCTTGTCGGGCCGCAT	Uncultured eubacterium Hb7-K3 (142-162 in AJ298577) partial 16S rRNA gene
20b	AAAGCGAGGTCGCTGCGCGTGACGCTGCCGAA	
	CGCTGCGCGTGACGCTGCCG	*Escherichia coli* (1836-1817 in X60999) rhsD gene (recombination hot spot)
	GCTGCGCGTGACGCTGCCG	*Serratia entomophila* plasmid pADAP (61066-61084 in AF135182) hypothetical protein
	CTGCGCGTGACGCTGCCG	*Ralstonia solanacearum* GMI1000 megaplasmid (62601-62584 in AL646078) probable lipoprotein transmembrane
21b	GCGAGCGACCGGCATGCCATCAGACCCGACCC	
	CGAGCGACCGGCATGCCAT	*Psychrobacter arcticum* 273-4 (2165270-2165252 in CP000082) probable amidase gene
	CGACCGGCATGCCATCA	*Pseudomonas putida* KT2440 (233443-233427 in AE016781) efflux membrane fusion protein

* in bold: mismatch.

DRB locus (2306022-)

1b
gtcttctccgcacacgcggaggtatttccacgtcggtgagtaaaactccgaaagttaccgggtcttctccgcacacgcgg

2b
aggtatttccccaaggaaatgtttcttaaactccccaccatcgtcttctccgcacacgcggaggtatttccaggtggtta

3b 4b
ccgtggcgaaccagtgcgctggggtcttctccgcacacgcggaggtatttccccgcggccgatgtgaccgcagcacgggg

5b
cgacgtcttctccgcacacgcggaggtatttccagtttcatatccctgaaataaactgttccagtggtcttctccgcaca

6b
cgcggaggtatttccatcgaagtaccggtctgccatgaggtcgttatagtcttctccgcacacgcggaggtatttccatt

7b 7b
ggtggataataggctaggtgattcttggtgtcttctccgcacacgcggaggtatttccattggtggataataggctaggt

8b
gattcttggtgtcttctccgcacacgcggaggtatttccgacgaaaaggaggcgatcgccggttctatcggtcttctcc

9b
gcacacgcggaggtatttccctacggctccacagggccaaaagttcgagaaagtcttctccgcacacgcggaggtatttc

10b 11b
caggcaagccgctggaggagcttactcgccaatgtcttctccgcacacgcggaggtatttccctgttttggttgtgcagt

12b
ccccgaacatatctgtcttctccgcacacgcggaggtatttccgatgaccgcggccttgtgcaggtcaagattaggtctt

13b
ctccgcacacgcggaggtatttccatggaaagaaatcgctgaacgccgactccgccgtcttctccgcacacgcggaggta

14b
tttccatacttaggatgtcccaaagctaaggtagcgggtcttctccgcacacgcggaggtatttcgcacgtaggcttgtc

15b 16b
gggccgcatatgaatggtgtcttctccgcacacgcggaggtatttccgtcttcgaggattcgacgggtttccattgtgcg

17b
tcttctccgcacacgcggaggtatttccaatcttgtgagggtcctcggatgaccgcgcaggtcttctccgcacacgcgga

18b
ggtatttccttaaagcaccatccacgggtgttatttgctgagtcttctccgcacacgcggaggtatttccagctcagatt

19b 20b
gcggggattctccccacatttgtcttctccgcacacgcggaggtatttctaaagcgaggtcgctgcgcgtgacgctgcc

17b
gaagtcttctccgcacacgcggaggtatttccaatcttgtgagggtcctcggatgaccgcgcaggtcttctccgcacacg

18b
cggaggtatttccttaaagcaccatccacgggtgttatttgctgagtcttctccgcacacgcggaggtatttccagctca

19b 20b
gattgcggggattctccccacatttgtcttctccgcacacgcggaggtatttctaaagcgaggtcgctgcgcgtgacgc

21b
tgccgaagtcttctccgcacacgcggaggtatttccgcgagcgaccggcatgccatcagacccgacccgtcttctccgca

cacgcggaggtatttcc (2306738)........

........(1.4 kb)..................hypothetical SSB protein gene..

(2307638) 22b
gtcttctccgcacacgcagaggtatttccaccgtcctcaccctcgtaggaccacaaggtattgtcttctccgcacacgtg

gaggtatttca (2309119)

Figure 3. DRB CRISPR locus in *Corynebacterium diphtheriae* NCTC13129, complete genome (Cerdeno-Tarraga et al., 2003), positions from 2306022 to 2309119 (in relation to *oriC*), accession number NC_002935. DRs are underlined. spacer number is given above the spacer sequence. Some spacers are duplicated. Region [DR - spacer 22b - DR] is separated from the main part of the DRB locus (~1.4 kb downstream).

In *C. diphtheriae*, five of the seven found homologies are on minus-strand sequences in GenBank which may also hint at the RNA interference as a mode of action in active CRISPR loci. Otherwise, a disrupted last DR in the *C. diphtheriae* DRA locus (only half present) may imply that DRA locus is no more active or at least cannot add more spacers/repeats in the downstream region. On the other hand, some homologies found for *C. diphtheriae* CRISPR spacers are within chromosomal genes (spacers 15b and 21b [Table 1]). Consequently, although a control of both mobility and maintenance of extrachromosomal genetic elements could be the most common function of CRISPRs, the homologies found with chromosomal sequences unrelated to foreign DNA suggest that CRISPRs may be involved in a more complex regulatory system.

CRISPR analysis also revealed striking points of similarity between *C. diphtheriae* and other *Corynebacterium* (*C. jeikeium*). The last spacer in the DRA locus (6a) showed homology to the direct repeat motif in the large 60-unit CRISPR locus in *C. jeikeium*. (Table 1). *C. jeikeium* is a skin pathogen and thus likely more ancient species than human respiratory pathogen *C. diphtheriae*. Accordingly, it seems more plausible that this spacer 6a was acquired from the *C. jeikeium* CRISPR locus and not vice versa. In order to understand origin and evolution of these loci in *C. diphtheriae*, we further studied regions flanking the DRA and DRB regions. A homology was found only for a short region immediately downstream to the DRA locus: to the orthologous IS (transposases) in *C. diphtheriae* and *C. jeikeium* genomes and also in *C. jeikeium* plasmid (Table 2). We compared the aligned part of these sequences, i.e., positions from +6 to +104 from the end of the last, abridged repeat in the DRA locus (Figure 4). Interestingly, branch lengths were very similar for orthologous IS sequences in both *Corynebacterium* species which demonstrated their true orthology and that they were already present in the common ancestor of both *C. diphtheriae* and *C. jeikeium* prior to their divergence. On the other hand, a very short branch for the DRA-downstream region in *C. diphtheriae* together with its 100% monophyly with *C. diphtheriae* IS sequence (Figure 4), suggests that the former was acquired from this IS sequence by intrachromosomal rearrangement within the *C. diphtheriae* genome but not from *C. jeikeium*. It is possible that this CRISPR region in *C. diphtheriae*, functionally or structurally, is prone to acquire DNA fragments, acting as "insertion preferential locus" (*ipl*) similar to such CRISPR (DR) locus in *M. tuberculosis* (Beggs et al. 2000; van Embden et al. 2000; Mokrousov et al. 2002). In the same line of analysis, the most recent study of DeBoy et al. (2006) have shown CRISPRs' association with large-scale intragenomic DNA rearrangements in hyperthermophiles *Thermotaga spp*.

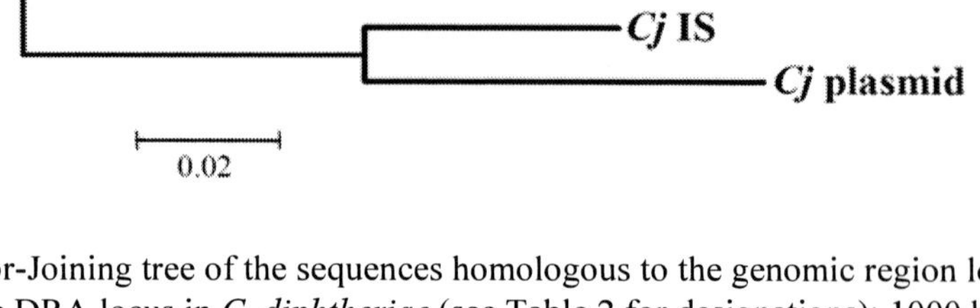

Figure 4. Neighbor-Joining tree of the sequences homologous to the genomic region located downstream to the DRA locus in *C. diphtheriae* (see Table 2 for designations); 1000 bootstrap replicates were used.

Conclusion

Recent studies demonstrated the other, practical facet of CRISPRs. In CRISPR loci, the order of motifs is always the same (van Embden et al. 2000; Pourcel et al. 2005). This fact together with a large number of particular spacers that may be present or absent in a locus, offer an abundance of variable characters and make these loci especially suitable for strain differentiation. The DNA reverse hybridization method was developed to study variation in the *M. tuberculosis* DR (CRISPR) locus (presence/absence of 43 different spacers) using the macroarray format; this method was named 'spoligotyping' (spacer oligonucleotide typing [Kamerbeek et al. 1997]) and has been widely used for epidemiological and phylogenetic purposes (Kamerbeek et al. 1997; Sola et al. 2001; Filliol et al. 2003; Brudey et al., 2006). The robustness of the spoligotyping based *M. tuberculosis* phylogeny was confirmed by other independent molecular markers (Filliol et al. 2006).

A computer program, CRISPRFinder, has very recently been developed which identifies CRISPRs and extracts the repeated and unique sequences (Grissa et al., 2007ab). Using this software, a database has been constructed which is automatically updated monthly from newly released genome sequences. Additional tools were created to allow the alignment of flanking sequences in search for similarities between different loci and to build dictionaries of unique sequences. To date, almost six hundred CRISPRs have been identified in 475 published genomes. Two Archeae out of thirty-seven and about half of Bacteria do not possess a CRISPR. Fine analysis of repeated sequences strongly supports the current view that new motifs are added at one end of the CRISPR adjacent to the putative promoter. Grissa et al. (2007a) hope that availability of a public database, regularly updated and which can be queried on the web will help in further dissecting and understanding CRISPR structure and flanking sequences evolution. Subsequent analyses of the intra-species CRISPR polymorphism will be facilitated by CRISPRFinder and the dictionary creator. CRISPRdb is accessible at http://crispr.upsud.fr/crispr.

Presumably because of strong structural constraints, the DR sequence is more conserved than the surrounding genomic elements, so that a CRISPR typing assay (in which a single PCR amplification with primers corresponding to the DR sequence is sufficient to amplify the whole set of spacers) might in theory be applied on a larger evolutionary scale than multi-locus VNTR analysis (Grissa et al., 2007ab). To date, CRISPR-based strain typing has been widely and systematically used only for *M. tuberculosis* (Kamerbeek et al. 1997; Sola et al. 2001). In few instances, interstrain variation in similar loci was shown in *Streptococcus* (Hoe et al. 1999; Bolotin et al. 2005) *Campylobacter* (Schouls et al. 2003; Price et al., 2007), and *Yersinia* (Pourcel et al. 2005). In our recent studies, we commenced development of the spoligotyping scheme for *C. diphtheriae* (Mokrousov et al. 2005, 2007). It appears that in the nearest future we will see an increasing use of CRISPRs for phylogenetical and epidemiological analyses of many bacterial species.

Table 2. Features in the sequences most similar to the CRISPR DRA locus downstream region in the *C. diphtheriae* genome

Query region	Sequence designation*	Positions in the query and level of identity with best match	Source, positions in the matching sequence, and accession number	Features in the matching sequence
DRA downstream (from position 39485)	*Cd* IS	**positions +6…127: 116/122 identities, no gaps	*Corynebacterium diphtheriae* gravis NCTC13129, positions 2080408-2080287, NC002935	78 bp at 5' side: Putative transposase for insertion sequence element (pseudogene); 71 bp at 3' side: Putative NAD(P)H-dependent reductase
	Cj IS	**positions +1…104: 87/104 identities, no gaps	*Corynebacterium jeikeium* K411, 2085049-2085152, CR931997	Transposase for IS*3525*
	Cj Plasmid	**positions +6…104: 82/99 identities, no gaps	*Corynebacterium jeikeium* plasmid pA501 (12196-12098, AY266269)	*ykwC* gene disrupted by transposon insertion
			Corynebacterium jeikeium plasmid pKW4 (9751-9653, AF401314)	*ykwC* gene - hypothetical 10.9 kDa protein; similar to N-terminal region of IS1201 transposase

*, here and in Figure 4.
**, position from the end of the last, abridged repeat in the DRA locus.

References

Brudey, K., Driscoll, J. R., Rigouts, L., Prodinger, W. M., Gori, A., Al-Hajoj, S. A., Allix, C., Aristimuno, L., Arora, J., Baumanis, V., Binder, L., Cafrune, P., Cataldi, A., Cheong, S., Diel, R., Ellermeier, C., Evans, J. T., Fauville-Dufaux, M., Ferdinand, S., Garcia de Viedma, D., Garzelli, C., Gazzola, L., Gomes, H. M., Gutierrez, M. C., Hawkey, P. M., van Helden, P. D., Kadival, G. V., Kreiswirth, B. N., Kremer, K., Kubin, M., Kulkarni, S. P., Liens, B., Lillebaek, T., Ly, H. M., Martin, C., Mokrousov, I., Narvskaya, O., Ngeow, Y. F., Naumann, L., Niemann, S., Parwati, I., Rahim, M. Z., Rasolofo-Razanamparany, V., Rasolonavalona, T., Rossetti, M. L., Rusch-Gerdes, S., Sajduda, A., Samper, S., Shemyakin, I., Singh, U. B., Somoskovi, A., Skuce, R., van Soolingen, D., Streicher, E. M., Suffys, P. N., Tortoli, E., Tracevska, T., Vincent, V., Victor, T. C., Warren, R., Yap, S. F., Zaman, K., Portaels, F., Rastogi, N., and Sola, C. (2006) *Mycobacterium tuberculosis* complex genetic diversity : mining the fourth international spoligotyping database (SpolDB4) for classification, population genetics and epidemiology. *BMC Microbiol.*, 6:23.

Beggs, M. L., Eisenach, K. D., and Cave, M. D. (2000) Mapping of IS*6110* insertion sites in two epidemic strains of *Mycobacterium tuberculosis, J Clin Microbiol*, 38, 2923-2928.

Benson G (1999) Tandem repeats finder: a program to analyze DNA sequences. *Nucleic Acids Res.*, 27:573-580.

Bolotin, A., Quinquis, B., Sorokin, A., and Ehrlich, S. D. (2005) Clustered regularly interspaced short palindrome repeats (CRISPRs) have spacers of extrachromosomal origin. *Microbiology, 151*, 2551-2561.

Cerdeño-Tarraga, A. M., Efstratiou, A., Dover, L. G., Holden, M. T. G., Pallen, M., Bentley, S. D., Besra, G. S., Churcher, C., James, K. D., De Zoysa, A., Chillingworth, T., Cronin, A., Dows, L., Feltwell, T., Hamlin, N., Holroyd, S., Jagels, K., Moule, S., Quail, M. A., Rabbinowitsch, E., Rutherford, K. M., Thomson, N. R., Unwin, L., Whitehead, S., Barrel, B. G., and Parkhill, J. (2003) The complete genome sequence and analysis of *Corynebacterium diphtheriae* NCTC13129. *Nucleic Acids Res.*, 31, 6516-6523.

DeBoy, R. T., Mongodin, E. F., Emerson, J. B., and Nelson, K. E. (2006) Chromosome evolution in the *Thermotagales*: large-scale inversions and strain diversification of CRISPR sequences. *J. Bacteriol.*, 188, 2364-2374.

Filliol, I., Driscoll, J. R., van Soolingen, D., Kreiswirth, B. N., Kremer, K., Valetudie, G., Dang, D. A., Barlow, R., Banerjee, D., Bifani, P. J., Brudey, K., Cataldi, A., Cooksey, R. C., Cousins, D. V., Dale, J. W., Dellagostin, O. A., Drobniewski, F., Engelmann, G., Ferdinand, S., Gascoyne-Binzi, D., Gordon, M., Gutierrez, M. C., Haas, W. H., Heersma, H., Kassa-Kelembho, E., Ho, M. L., Makristathis, A., Mammina, C., Martin, G., Moström, P., Mokrousov, I., Narbonne, V., Narvskaya, O., Nastasi, A., Niobe-Eyangoh, S. N., Pape, J. W., Rasolofo-Razanamparany, V., Ridell, M., Rossetti, M. L., Stauffer, F., Suffys, P. N., Takiff, H., Texier-Maugein, J., Vincent, V., de Waard, J. H., Sola, C., and Rastogi, N. (2003) Snapshot of moving and expanding clones of *Mycobacterium tuberculosis* and their global distribution assessed by spoligotyping in an international study. *J. Clin. Microbiol.*, 41, 1963-1970.

Filliol, I., Motiwala, A. S., Cavatore, M., Qi, W., Hernando Hazbon, M., Bobadilla Del Valle, M., Fyfe, J., Garcia-Garcia, L., Rastogi, N., Sola, C., Zozio, T., Guerrero, M. I., Leon, C. I., Crabtree, J., Angiuoli, S., Eisenach, K. D., Durmaz, R., Joloba, M. L., Rendon, A., Sifuentes-Osornio, J., Ponce de Leon, A., Cave, M. D., Fleischmann, R., Whittam, T. S., and Alland, D. (2006) Global phylogeny of *Mycobacterium tuberculosis* based on single nucleotide polymorphism (SNP) analysis: insights into tuberculosis evolution, phylogenetic accuracy of other DNA fingerprinting systems, and recommendations for a minimal standard SNP set. *J. Bacteriol.*, 188, 759-772.

Grissa, I., Bouchon, P., Pourcel, C. and Vergnaud, G. (2007) On-line resources for bacterial micro-evolution studies using MLVA or CRISPR typing. *Biochimie,* in press.

Grissa, I., Vergnaud, G. and Pourcel, C. (2007) The CRISPRdb database and tools to display CRISPRs and to generate dictionaries of spacers and repeats. *BMC Bioinformatics, 8,* 172.

Hoe, N., Nakashima, K., Grigsby, D., Pan, X., Dou, S. J., Naidich, S., Garcia, M., Kahn, E., Bergmire-Sweat, D., and Musser, J. M. (1999) Rapid molecular genetic subtyping of serotype M1 group A *Streptococcus* strains. *Emerg. Infect. Dis.*, 5, 254-263.

Jansen, R., van Embden, J. D. A., Gaastra, W., and Schouls, L. M. (2002) Identification of genes that are associated with DNA repeats in prokaryotes. *Mol. Microbiol.*, 43, 1565-1575.

Kamerbeek, J., Schouls, L., Kolk, A., van Agterveld, M., van Soolingen, D., Kuijper, S., Bunschoten, A., Molhuizen, H., Shaw, R., Goyal, M., and van Embden, J. D. A. (1997) Simultaneous detection and strain differentiation of *Mycobacterium tuberculosis* for diagnosis and epidemiology. *J. Clin. Microbiol.*, 35, 907-914.

Kumar, S., Tamura, K., Jakobsen, I. B., and Nei, M. (2001) MEGA2: molecular evolutionary genetics analysis software. *Bioinformatics*, 17, 1244-1245.

Mojica. F. J., Diez-Villasenor, C., Garcia-Martinez, J., and Soria, E. (2005) Intervening sequences of regularly spaced prokaryotic repeats derive from foreign genetic elements. *J. Mol. Evol.*, 60, 174-182.

Mojica, F. J., Diaz-Villasenor, C., Soria, E., and Juez, G. (2000) Biological significance of a family of regularly spaced repeats in the genomes of Archaea, Bacteria and mitochondria. *Mol. Microbiol.*, 36, 244-246.

Mokrousov, I., Limeschenko, E., Vyazvaya, A. and Narvskaya, O. (2007) *Corynebacterium diphtheriae* spoligotyping based on combined use of two CRISPR loci. *Biotechnol. J.*, 2, 901-906.

Mokrousov, I., Narvskaya, O., Limeschenko, E., Otten, T., and Vyshnevskyi, B. (2002) Novel IS*6110* insertion sites in the direct repeat locus of *Mycobacterium tuberculosis* clinical strains from the St. Petersburg area of Russia, and evolutionary and epidemiological considerations. *J. Clin. Microbiol.*, 40, 1504-1507.

Mokrousov, I., Narvskaya, O., Limeschenko, E., and Vyazovaya, A. (2005) Efficient discrimination within a *Corynebacterium diphtheriae* epidemic clonal group by a novel macroarray-based method. *J. Clin. Microbiol.*, 43, 1662-1668.

Popovic, T., Mazurova, I. K., Efstratiou, A., Vuopio-Varkila, J., Reeves, M. W., De Zoysa, A., Glushkevich, T., and Grimont, P. (2000) Molecular epidemiology of diphtheria. *J. Infect. Dis.*, 181, S168-S177.

Pourcel, C., Salvignol, G., and Vergnaud, G. (2005) CRISPR elements in *Yersinia pestis* acquire new repeats by preferential uptake of bacteriophage DNA, and provide additional tools for evolutionary studies. *Microbiology*, 151, 653-663.

Price, E. P., Smith, H., Huygens, F., and Giffard, P. M. (2007) High-resolution DNA melt curve analysis of the clustered regularly interspaced short palindromic repeat locus of *Campilobacter jejuni*. *Appl. Environ. Microbiol.*, *73*, 3431-3436.

Saunders, N. F., Goodchild, A., Raftery, M., Guilhaus, M., Curmi, P. M., and Cavicchioli, R. (2005) Predicted roles for hypothetical proteins in the low-temperature expressed proteome of the Antarctic archaeon *Methanococcoides burtonii*. *J. Proteome Res.*, 4, 464-472.

Schouls, L. M., Reulen, S., Duim, B., Wagenaar, J.A., Willems, R. J., Dingle, K. E., Colles, F. M., and Van Embden, J. D. (2003) Comparative genotyping of *Campylobacter jejuni* by amplified fragment length polymorphism, multilocus sequence typing, and short repeat sequencing: strain diversity, host range, and recombination. *J. Clin. Microbiol.*, 41, 15-26.

Sola, C., Filliol, I., Legrand, E., Mokrousov, I., and Rastogi, N. (2001) *Mycobacterium. tuberculosis* phylogeny reconstruction based on combined numerical analysis with IS*1081*, IS*6110*, VNTR and DR-based spoligotyping suggests the existence of two new phylogeographical clades. *J. Mol. Evol.*, 53, 680-689.

Tang, T. H., Bachellerie, J. P., Rozhdestvensky, T., Bortolin, M. L., Huber, H., Drungowski, M., Elge, T., Brosius, J., and Hüttenhofer, A. (2002) Identification of 86 candidates for small non-messenger RNAs from the archaeon *Archaeoglobus fulgidus*. *Proc. Natl. Acad .Sci. USA*, 99, 7536-7541.

van Embden, J. D. A., van Gorkom, T., Kremer, K., Jansen, T., van der Zeijst, B. A. M., and Schouls, L. M. (2000) Genetic variation and evolutionary origin of the direct repeat locus of *Mycobacterium tuberculosis* complex bacteria. *J. Bacteriol.*, 182, 2393-2401.

Vitek C. R., and Wharton, M. (1998) Diphtheria in the former Soviet Union: reemergence of a pandemic disease. *Emerg. Infect. Dis.*, 4, 539-549.

Index

B

C

D

E

F

G

H

I

J

K

L

M

N

O

P

Q

R

S

T

U

V

W

X

Y

Z